New Students and New Places

POLICIES FOR THE FUTURE GROWTH AND DEVELOPMENT OF AMERICAN HIGHER EDUCATION

•A15042 000477

A Report and Recommendations by

The Carnegie Commission on Higher Education

OCTOBER 1971

MCGRAW-HILL BOOK COMPANY

New York St. Louis San Francisco Düsseldorf
London Sydney Toronto Mexico Panama
Johannesburg Kuala Lumpur Montreal
New Delhi Rio de Janeiro Singapore

*This report is issued by the Carnegie Commission on
Higher Education, with headquarters at
1947 Center Street, Berkeley, California 94704.
The views and conclusions expressed in this report
are solely those of the members of the Carnegie Commission
on Higher Education and do not necessarily reflect the
views or opinions of the Carnegie Corporation of New York,
The Carnegie Foundation for the Advancement of Teaching,
or their trustees, officers, directors, or employees.*

Copyright © 1971 by The Carnegie Foundation for
the Advancement of Teaching. All rights reserved.
Printed in the United States of America.

*Additional copies of this report may be ordered from
McGraw-Hill Book Company, Hightstown, New Jersey 08520.
The price is $3.50 a copy.*

Contents

Foreword, v

1 *Major Themes,* 1

2 *American Higher Education Today,* 11
Explosive growth ▪ Enrollment rates ▪ The wide variety of institutions ▪ Race ▪ Socioeconomic status ▪ Sex ▪ Regional diversity ▪ Metropolitan and nonmetropolitan areas ▪ International comparisons of enrollment rates

3 *The Future of Higher Education: The Questions to Be Examined,* 39

4 *The Future of Higher Education: Assumption A—Largely Uninhibited Growth,* 41
Enrollment increases ▪ Enrollment changes by type of institution

5 *The Future of Higher Education: Assumption B—Carnegie Commission Goals and Other Influences,* 49
Constructive change ▪ Alternative enrollment estimates ▪ Cost estimates

6 *The Growth of Institutions,* 65
Optimum size ▪ Cluster colleges ▪ Federations and consortia ▪ Preserving and encouraging diversity

7 *Needs for New Institutions,* 97
Introduction ▪ Needs for new urban institutions ▪ Other needs for new institutions

8 *Toward More Flexible Patterns of Participation in Higher Education,* 111
The need for reform in traditional types of adult higher education ▪ External degrees and The Open University in Britain ▪ Similar developments in the United States

9 *Summary,* 119

Appendix A: Carnegie Commission Classification of Institutions of Higher Education, 1970, 121

Appendix B: Tables, 127

References, 155

Foreword

This report, *New Students and New Places: Policies for the Future Growth and Development of American Higher Education,* presents the Commission's projections of enrollment in higher education to the year 2000 and develops estimates of the ways in which these enrollment projections would be affected by implementation of the recommendations included in the Commission's earlier reports and in this report. Also presented for the first time are enrollment data and projections based on the Commission's classification of institutions of higher education. We plan to publish the list of institutions as classified by the Commission within the next few months.

In addition, the report includes policy recommendations relating to (1) the growth of institutions, (2) maintaining innovation and diversity in higher education, (3) needs for new institutions, and (4) encouraging more flexible patterns of participation in higher education.

To the many persons who were consulted and gave us helpful suggestions, we wish to express our appreciation. Particularly valuable contributions were made by Dr. John K. Folger, executive director, Tennessee Higher Education Commission, who participated in the development of the Commission's classification of institutions and in the preparation of the estimates of needs for new institutions. The enrollment projections were developed under the direction of Dr. Gus W. Haggstrom of the University of California, Berkeley.

We also wish to thank the members of our staff, and especially Dr. Margaret S. Gordon, for their work in preparing this report.

Eric Ashby
The Master
Clare College
Cambridge, England

Ralph M. Besse
Chairman of the Board
National Machinery Company

Joseph P. Cosand
Director
Center for Higher Education
University of Michigan

William Friday
President
University of North Carolina

Patricia Roberts Harris
Partner
Fried, Frank, Harris, Shriver,
  & Kampelman

David D. Henry
Distinguished Professor of Higher
  Education
University of Illinois

Theodore M. Hesburgh, C.S.C.
President
University of Notre Dame

Stanley J. Heywood
President
Eastern Montana College

Carl Kaysen
Director
Institute for Advanced Study
  at Princeton

Kenneth Keniston
Professor of Psychology
Yale University School of
  Medicine

Katharine E. McBride
President Emeritus
Bryn Mawr College

James A. Perkins
Chairman of the Board
International Council for
  Educational Development

Clifton W. Phalen
Chairman of the Executive
  Committee
Marine Midland Banks, Inc.

Nathan M. Pusey
President
The Andrew W. Mellon Foundation

David Riesman
Professor of Social Relations
Harvard University

The Honorable William Scranton
Chairman
National Liberty Corporation

Norton Simon

Kenneth Tollett
Distinguished Professor of Higher
  Education
Howard University

Clark Kerr
Chairman

New Students
and New Places

1. Major Themes

1 Higher education in the United States has comprised a continuous rapid growth segment of the nation for more than three centuries. During that time, it has experienced steady enrollment increases at a rate faster than the expansion of American society generally. Over the past century, in particular, enrollments in higher education have doubled regularly every 14 to 15 years.[1] But never again.

2 A new period, marking a change in growth dynamics of great historic importance, is now beginning.

Decade	Changes in enrollments
1960–1970	Doubled
1970–1980	Increase by one-half
1980–1990	None
1990–2000	Increase by one-third

Percentage change in enrollments in accordance with past trends

Period	%
1960–1970	124
1970–1980	59
1980–1990	−1
1990–2000	30

[1] Eric Ashby, *Any Person, Any Study: An Essay on Higher Education in the United States*, first of a series of essays sponsored by The Carnegie Commission on Higher Education, McGraw-Hill Book Company, New York, 1971, p. 4.

The new period is marked by two features:

Go-Stop-Go in the short run The Stop period of the 1980s will bring new construction to a halt, greatly slow the recruitment of new faculty members, and reduce opportunities for reform (which are more likely to arise during periods of expansion when new campuses are being created and established ones are starting new endeavors). But it will also facilitate some improvements in quality since the demands of greater quantity will have disappeared. With less attention being given to accommodating additional college-age youth, expansion in the education of adults may be additionally undertaken on a major scale. This Stop period will be preceded by a Go period in the 1970s and followed by another Go period in the 1990s.

Reduced growth rate in the long run Higher education historically has grown much faster than society as a whole because of rapidly increasing attendance rates (from about 2 percent of the college-age group in 1870 to about 35 percent in 1970). This particular growth rate will approach a limit. We now expect the percentage of college-age population actually in college at any moment of time to level off at about 50 percent[2] in the year 2000, although there are many uncertainties about its ultimate resting point. The 50 percent level will make it as possible for young persons of equal ability from the lower half of the socioeconomic scale to attend higher education as it is now for young persons from the upper half, and for all states to reach an average at the present level of the highest states in terms of high school graduation rate (90 percent) and entry of high school graduates into college (75 percent).[3] Higher attendance will also come from among persons now considered above college age. But formal higher education, even with the somewhat higher attendance rates among college-age youth and more adult enrollments, will be growing more nearly with society than rapidly ahead of it.

Higher education has never before faced the twin impacts of such a Go-Stop-Go situation and the cessation of a long-run growth rate substantially beyond that of American society.

[2] With perhaps two-thirds *entering* college, including those who do not remain very long. Thus a 66.7 percent entrance rate will mean about a 50 percent rate of attendance of young people between the ages of 18 and 21.

[3] Ninety percent of the age group graduating from high school and 75 percent of those entering college yield about two-thirds of the age group entering college.

3 Predictions of future enrollments, however, are now becoming more uncertain. Looking backward, what happened was predictable—a steady increase. Looking forward, there are these new uncertainties:

Financial stringency—at least through the 1970s and as great as ever in history. Will there be the new places created for the additional prospective students?

Labor market conditions—80 percent of jobs do not require a college degree, yet 65 percent of young people may soon at least enter higher education. What will happen when the labor market no longer generally absorbs college graduates at the level of training they have acquired, as is already happening in restricted fields?

The cultural revolution—a great unknown. More young people are now seeking vocations or life-styles outside the Horatio Alger syndrome than ever before. How far will this go? Higher education is built on the work ethic and we may now be shifting to a more sensate culture.

The birthrate—now fluctuates enormously as compared with historic standards. Recently it has gone down dramatically. Will it keep going down, or stabilize, or rise again? This factor alone makes predictions after 1990 quite risky. Now that births can be better controlled, the birthrate is less predictable.

The loosening of educational structures—one clear change now going on is to allow students to stop out more from education after high school and adults to stop in more than ever before.[4] How will the stop-outs and stop-ins balance each other?

The new technology—slides, tapes, computers, video cassettes. These all allow education to be greatly dispersed, bringing the best and the most recent knowledge into what were once remote and isolated colleges, into factories and offices, and into houses. Every living room can soon be a classroom. This will certainly dispense higher education to many more people than ever before; and it will also disperse it over many more locations. The campus may no longer be the sole, or possibly even dominant, location for higher education.

Public policy—now in a period of reassessment after one century of strong support for higher education. How much aid will there

---

[4] The percentage of the population between the ages of 25 and 35 enrolled in higher education has tripled between 1950 and 1970.

be in the future for students and how much for institutions, with obvious impacts on growth?

Never in the history of higher education in the United States have so many uncertainties surrounded the future course of growth.

4 We predict, nevertheless, the following enrollments, first on the basis of past and current trends and, second, on the basis of prospective trends reflecting new policies and developments:

Past and current trends:

Year	Total enrollments
1970	8,500,000
1980	13,500,000
1990	13,300,000
2000	17,400,000

Prospective trends:

Year	Total enrollments
1970	8,500,000
1980	12,500,000
1990	12,300,000
2000	16,000,000

The "prospective trends" are, in reality, the trends we would like to see develop on the basis of recommendations of the Carnegie Commission.

This "alternative future" will be different from the past and current trends as follows:

More students as a result of —greater equality of opportunity

—more adult education

Fewer students as a result of —reduced time per degree

—shift of enrollments to two-year colleges

—reduced graduate enrollment

The net effect of these changes would be a reduction of "status quo" estimates of students in 1980 in the general order of 1 million. (See Table 4.) We believe that higher education would be in a healthier state in 1980 with about 1 million fewer students than current trends indicate, and these fewer students would be more equitably drawn from the total population if our policy recommendations were to become effective practice.

Thirty years hence, in the year 2000, regardless of which projections are used, higher education will be roughly 10 times larger in terms of enrollments than it was 30 years ago in 1940.

5   The additional new students could be mostly absorbed in 1980 (and thus also in 1990) within the existing 2,800 campuses in the United States without either forcing undue size or unduly rapid growth on these campuses—with one very major qualification: these campuses are not uniformly well located for this purpose.

We find a major deficit in two types of institutions—community colleges and comprehensive colleges—in metropolitan areas, especially those with a population over 500,000. The inner cities, in particular, are not now well served by higher education. Higher education has not adequately reflected the urbanization of America. Deficits in North Jersey and the eastern side of Chicago are illustrative. Thus we recommend additional campuses by 1980 as follows:

Community colleges	175–235
Comprehensive colleges	80–105

We find no need whatsoever in the foreseeable future for any more research-type universities granting the Ph.D. Available resources should be concentrated on those that now exist rather than on creating new ones.

We hope very much that the private liberal arts colleges can maintain their position within the universe of American higher education. They provide diversity, innovative possibilities, competitive pressure on public institutions to serve well their individual students, and standards of institutional autonomy. Some will inevitably merge with other colleges or disappear; but we trust that new institutions may come to take their places. We recommend policies that would lead to special federal and state assistance to these colleges.

6   We are concerned about the size of individual campuses as well as with the totality of higher education. We are convinced that some institutions are too small to be effective either in the use of their resources or in the breadth of the program they offer their

students—the "cult of intimacy" has its academic limits; a "critical mass" is necessary for successful educational endeavors. Thus we suggest certain "peril points" below which individual institutions should examine their special histories, their locations, and their purposes and philosophies to determine whether their small size really is desirable. Our suggested peril points (on the basis of full-time equivalent enrollment) are:

For liberal arts colleges	1,000
For community colleges	2,000
For comprehensive colleges	5,000
For universuty campuses	5,000

We are also concerned that some institutions engage in mindless growth and subscribe to a "cult of gigantism"—that bigger is always better; and that many state finance agencies assume that bigger is at least cheaper. We find this not to be the case beyond relatively modest size. We also find no evidence that academic quality necessarily increases beyond a modest size. We ask that there be consideration for the costs of size as well as the advantages. Such costs include:

Loss of personal attention to students

Loss of personal acquaintance among faculty members

Increase in administrative complexity

Increase in disruptive events on campus

Loss of the chance to serve new areas with new campuses

Loss of the chance to diversify with new and different types of campuses

Thus we suggest that institutions examine carefully their growth plans as they exceed certain limits—"points for reassessment," again realizing that there is great dissimilarity among institutions in their histories, their locations, their purposes, their political realities. Our suggested points of reassessment (on the basis of full-time equivalent enrollments) are:

For liberal arts colleges	2,500
For community colleges	5,000
For comprehensive colleges	10,000
For university campuses	20,000

Many institutions now lie outside these lower and upper limits beyond which special consideration of optimum size is suggested.

	Approximate percentages of institutions with FTE enrollments below our suggested minima or above our suggested maxima	
	Below minima	Above maxima
Doctoral-granting institutions		
Public	3	23
Private	32	6
Comprehensive colleges		
Public	58	12
Private	90	0
Liberal arts colleges		
Public	74	7
Private	68	1
Two-year institutions		
Public	70	8
Private	98	0

We recognize, in particular, that certain institutions with carefully designed curricula (like Shimer College, with its core curriculum) or with a limited range of programs (like Princeton, with few graduate professional schools, and Cal Tech and Rockefeller University, with a carefully chosen spread of academic fields) or with successful consortia arrangements (like the Claremont group of colleges) may be more effective at a smaller size than if they had more customary curricula or a broader range of programs or a more self-contained approach. We also recognize that the problems of large size can be mitigated by internal decentralization into "cluster colleges," semiautonomous professional schools, or separately identifiable "theme colleges" (like Harvard, Yale, Santa Cruz, Wisconsin—Green Bay); and we recommend strongly consideration of such decentralization within institutions of large scale to reduce the "curse of bigness."

There are "golden means" in the sizes and internal structures of campuses, as in so many other human endeavors. Traditionally in the United States we have given little consideration to either the optimum size of a campus or its effective internal organization, accepting unlimited growth and historically given internal structures. The British, by comparison, intensely have debated both issues.

There are also "golden mean" growth rates for campuses. Unduly rapid growth makes it difficult to select and absorb new faculty members, select and train and season new administrators, select and guide new endeavors. At the same time, the sudden cessation of rapid growth can also be a traumatic experience, with horizons suddenly limited and confidence in the future shaken. Thus we suggest that modest growth rates long sustained are better than rapid rates suddenly cut off. We realize, of course, that percentage growth rates must, of necessity, be very rapid when a campus is first started. We suggest, however, that, under normal circumstances, growth rates of 3 to 5 percent a year not be exceeded. They make possible, on a cumulative basis, growth of over 50 percent in a single decade. These growth rate considerations are relevant to the 1970s and 1990s but not to the 1980s.

Generally, we suggest that much more attention be given than has usually been the case in the past to:

Effective size

Suitable internal structure

Desirable rate of growth

7 The Carnegie Commission recommendations relating to the growth of higher education generally are mainly these:

Increasing equality of opportunity, particularly for students whose families are in the lower half of the socioeconomic scale—which includes many minority families

Improving the location of campuses by placing a community college within commuting distance of 95 percent of all Americans and adding comprehensive colleges particularly in the inner cities of metropolitan areas

Reducing the length of time it takes for students to earn degrees

Loosening educational structures and rules so that young persons can stop out of education and adults can enter more readily

Dispersing opportunities beyond the conventional campus through "open universities," external degree programs, video cassettes, and other new institutional devices and new technology

Preserving and even increasing the diversity of institutions of higher education by type and by program; resisting homogenization

Holding steady the number of universities

Preserving the private sector of higher education in general and the liberal arts colleges in particular

Considering more thoughtfully the size, the internal structure, and the rate of growth of individual campuses

Reducing costs in the process of adapting higher education more effectively to the needs of students and of American society

8 Higher education in the United States until about 1940 was largely for the elite; from 1940 to 1970, we moved to mass higher education; and, from 1970 to 2000, we will move to universal-access higher education—opening it to more elements of society than ever before. We do not anticipate a further move to universal higher education in the sense of universal attendance; in fact, we consider this undesirable and believe that public and private policy should both avoid channeling all youth into higher education and create attractive alternatives to higher education. But we clearly are moving from mass to universal-access higher education. This creates problems. It also creates opportunities for more nearly equal treatment of all our citizens, for more nearly adequate service to all localities of our nation, for more varied responses to the increasingly varied composition of the enrollments in higher education, for new methods and new types of institutions, for a more thoughtful consideration of the future role of each of the major components of our universe of higher education, for a more careful look at the essential nature of each of our institutions, and for a more systematic examination of the effective use of resources.

9 The United States is creating a society in which more people will have had more education than ever before in history in any nation. Thus it is encountering new problems and opportunities for itself and charting new territory for other nations. With a few major exceptions, particularly Canada, whose situation roughly corresponds to that in the United States, most other nations of the world are now beginning to move or are completing the move from a historically more elite system to mass higher education to meet the technical requirements of industrial society. The United States is now going beyond those requirements and making higher education available to all who want it for whatever reason. It is reaching the stage of "any person, any study."

2. American Higher Education Today

EXPLOSIVE GROWTH

The 1960s were characterized by explosive growth in American higher education. Total enrollment in all institutions reporting to the U.S. Office of Education rose from 3.8 million in 1960 to 8.5 million in 1970. The numerical increase of 4.7 million students was by far the largest ever experienced in one decade, while in percentage terms the increase (124 percent) was comparable to the largest 10-year increase previously recorded, from 1870 to 1880 (Appendix B, Table 2). Changes in degree-credit enrollment from 1930 to 1970, with estimates of future changes to the year 2000, are presented in Chart 1.

In addition to the 8.5 million enrolled in institutions reporting to the Office of Education, about 1.5 million students are enrolled in vocationally oriented schools of a largely proprietary nature. These schools are not included in Office of Education enrollment surveys because their programs tend to be nonaccredited.

What accounted for the extraordinarily rapid increase in enrollment in higher education in the 1960s? A major factor was the enrollment during this decade of young people who were born in the relatively high birthrate years of World War II and the early postwar period (Appendix B, Table 3). A baby born in 1942, when the birthrate rose quite sharply, would have been ready to enter college in 1959 or 1960, on the average. A child born in 1947, when the birthrate reached its postwar peak, would have been ready to enter college about 1964 or 1965.

All told, the rise in the college-age population accounted for about 45 percent of the increase in undergraduate degree-credit enrollment from 1960 to 1970, while the rest of the increase was attributable to a rise in the enrollment rate—i.e., the ratio of undergraduate degree-credit enrollment to the population aged 18 to 21—from 33.8 percent in 1960 to an estimated 47.5 percent in 1970 (Appendix B, Table 1).

CHART 1 Total degree-credit enrollment in institutions of higher education, actual, 1930-1970, and projection, 1971-2000 (numbers in thousands—logarithmic scale)

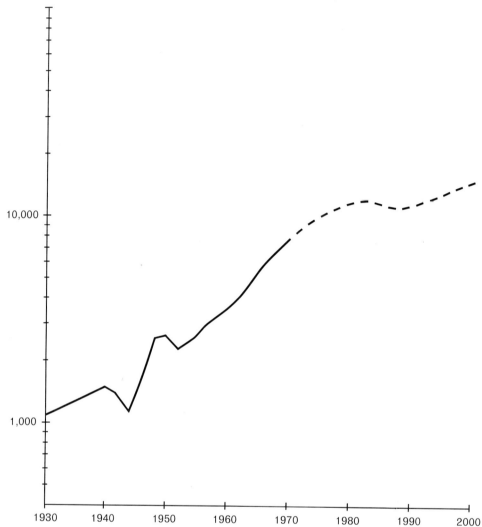

NOTE: Future enrollments are based on the Carnegie Commission's Projection C. The assumptions underlying three projections of the Commission, A, B, and C, are discussed briefly in Section 4.
SOURCE: U.S. Office of Education and projections developed by Carnegie Commission staff.

Looking back over the decades we find that the enrollment rate rose quite slowly and gradually until about 1945 and then began to rise at an accelerated rate. The high rates of the years immediately following World War II were, of course, associated with heavy enrollment of veterans, but a second sharp increase in the enrollment rate began following the Korean conflict and continued unabated throughout the remainder of the 1950s and the entire decade of the 1960s. Martin Trow has compared the increase from about 15 percent in 1940 to nearly 50 percent in 1970 to the very similar burgeoning of high school enrollment rates between 1910 and 1940 (1, p. 152).*

In the fall of 1970, nearly three-fifths of all students enrolled were men and about two-fifths were women. Approximately seven-tenths were enrolled full time and the remainder on a part-time basis. This meant that full-time equivalent enrollment amounted to about 6.7 million students.

The upward trend in graduate enrollment in the 1960s was even more spectacular than the increase in undergraduate enrollment. In 1960 there were about 356,000 graduate resident students enrolled in master's and doctor's degree programs. Our projection indicates that the corresponding figure for 1970 was about 930,000 (Appendix B, Table 2), representing an increase of 161 percent in the 1960s. In fact, graduate enrollment has been rising at a more rapid rate than undergraduate enrollment throughout the present century.

ENROLLMENT RATES

Let us look more closely at the behavior of enrollment rates. Undergraduate degree-credit enrollment includes persons who are younger than age 18 and older than age 21, while total degree-credit enrollment also includes persons who are not within the 18-to-24 age range. To determine the actual percentage of the college-age population enrolled, we must turn to U.S. Bureau of the Census data, available only since 1940 (Chart 2). Between 1940 and 1970, the proportion of persons aged 18 to 21 enrolled in college rose from 11 to 34 percent, and we project an increase to about 54 percent by the year 2000. As for the 18-to-24 age group, 26 percent were enrolled in 1970, as compared with 8 percent in 1940. This proportion is likely to rise to about 41 percent by 2000.

The actual percentages of the college-age population enrolled, then, are considerably below the enrollment rates based on Office of Education total degree-credit enrollment of persons of all ages as a percentage of the college-age population (Chart 3). Significantly, also, the difference between the two measures of enrollment rates widened considerably from 1940 to 1970, particularly as

---

*See "References" section for complete citations.

CHART 2 *Percentage of the college-age population enrolled in degree-credit programs in higher education, 1940 to 1970, and projected, 1980 to 2000*

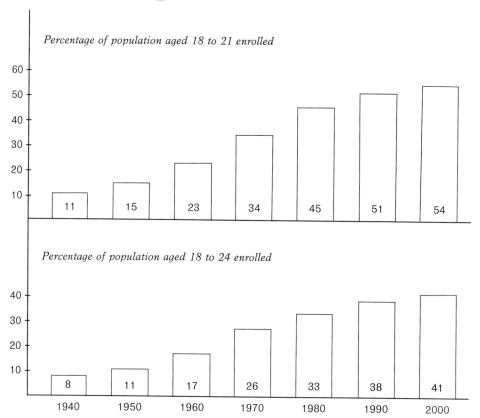

SOURCE: U.S. Bureau of the Census and projections developed by Carnegie Commission staff.

related to the population of 18- to 21-year-olds. By 1970, undergraduate degree-credit enrollment represented 48 percent of the population aged 18 to 21, whereas only 34 percent of this age group was enrolled. This means, of course, that the age range of undergraduates has widened considerably.

Both the Census and the Office of Education data presented in Charts 2 and 3 relate to degree-credit enrollment. Only in relatively recent years has the Office of Education collected statistics on non-degree-credit enrollment, a rapidly growing component of total enrollment in institutions of higher education. From 1960 to 1970, non-degree-credit enrollment increased from 206,000 to 720,000, or about 250 percent. Most of this increase occurred in two-year institutions, which accounted for nearly 90 percent of total non-degree-credit enrollment toward the end of the decade.

For the most part, in the remainder of this report we shall be using data on *total enrollment,* including both degree-credit and non-degree-credit enrollment. It is only in connection with historical series, or with comparisons between Census and Office of Education data, that it is necessary to rely on degree-credit enrollment.

THE WIDE VARIETY OF INSTITUTIONS College and university students were enrolled in more than 2,800 institutions of higher education in 1970—the product of 334 years of growth and development of American colleges and universities from the founding of Harvard College in 1636. Private institutions dominated the early development of American higher education,

CHART 3 *Degree-credit enrollment as percentage of college-age population, 1940 to 1970, and projected, 1980 to 2000*

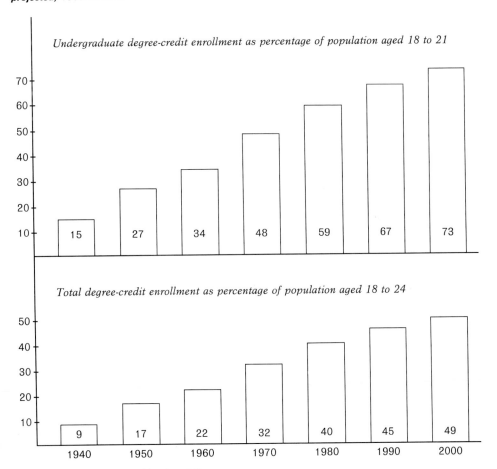

SOURCE: Appendix B, Tables 1 and 2.

but by 1970, three-fourths of all students were enrolled in publicly controlled institutions and all indications pointed to a continuation of the upward trend in the proportion enrolled in public institutions which characterized the first two decades of the century and the period since World War II (Chart 4).

Among private institutions of higher education in 1968, the majority were controlled by religious bodies, but less than half of the students were enrolled in these institutions.

Private institutions of higher education and their enrollment, by control, 1968

	Institutions	Enrollment
Number	1,529	2,100,900
Percent		
Proprietary	2.2	1.4
Independent	40.8	54.2
Catholic	22.6	19.7
Protestant	32.2	22.4
Other religions	2.2	2.3
TOTAL	100.0	100.0

SOURCE: Adapted from U.S. Office of Education data by the Carnegie Commission staff.

More than two-fifths of the private institutions were under independent auspices, and these colleges and universities enrolled 54 percent of the students in private higher education. Next most numerous were Protestant institutions, accounting for nearly one-third of the private institutions and for slightly more than one-fifth of the students. Catholic institutions represented somewhat more than one-fifth of the private institutions and enrolled slightly less than one-fifth of the students.[1]

Within both the public and private sectors of American higher education there is an extraordinarily wide variety of institutions. The Carnegie Commission has developed a new classification of institutions of higher education designed to provide more meaningful and homogeneous categories than are found in other existing classifications. A description of these categories is included in Appendix A. Our classification includes 5 major groups of institutions and 17 subgroups.

---

[1] Studies of Catholic higher education (2) and of independent and Protestant liberal arts colleges (3) have been published under the auspices of the Commission.

It should be explained in this connection that, when we refer to an institution in this report, *we are referring in nearly all instances to a campus* (see Appendix A). The 2,827 institutions reporting to the Office of Education in 1970 were campuses, except in a few instances in which one institution reported enrollment for an entire system of campuses. This occurred, for example, in some of the

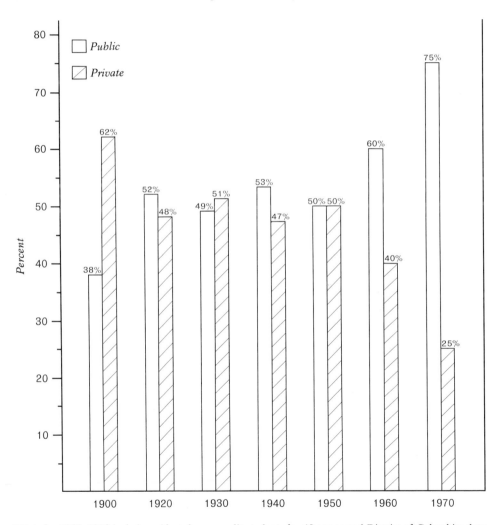

CHART 4 Enrollment in institutions of higher education, by control, United States, 1900–1970*

* Data for 1900–1950 include resident degree-credit students for 48 states and District of Columbia; data for 1960 and 1970 include total resident and extension students for 50 states and the District of Columbia.
SOURCE: U.S. Office of Education.

systems of two-year branch campuses of universities, as in Kentucky.

Doctoral-granting institutions Of the 2,827 campuses reporting to the Office of Education in 1970, 164, or less than 6 percent, were doctoral-granting institutions (Chart 5 and Appendix B, Table 4). They enrolled 30 percent of all students and ranged from cam-

CHART 5 Enrollment in institutions of higher education, and number of institutions, by type, 1963 and 1970

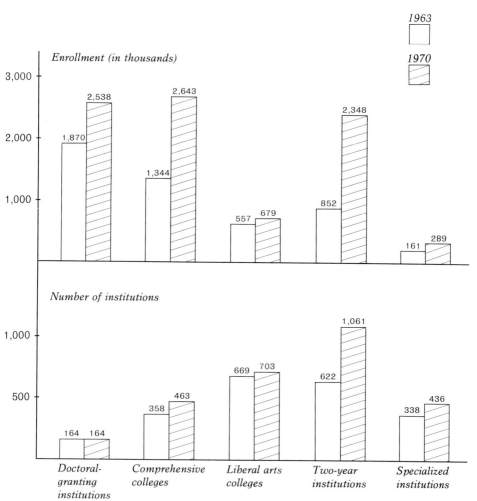

SOURCE: Adapted from U.S. Office of Education data by Carnegie Commission staff.

puses with heavy emphasis on research and outstanding national and international reputations to less prestigious campuses with modest research and doctoral-training programs. More than 100, or 62 percent, of these campuses were public, and they enrolled about three-fourths of the students in doctoral-granting institutions.

The most distinctive characteristic of the doctoral-granting institutions is the relatively large representation of graduate students in their enrollment. About one-fourth of all the students enrolled in these institutions in 1968 were in postbaccalaureate programs (Appendix B, Table 5), but postbaccalaureate students accounted for a larger proportion (37 percent) of all students in the private institutions than in the public institutions (21 percent).[2] Even so, the *number* of postbaccalaureate students in the public institutions (376,000) exceeded by a considerable margin the number in private institutions (233,000). It was not more extensive postbaccalaureate programs in the private institutions that accounted for the greater preponderance of postbaccalaureate enrollments within their student bodies, but relatively selective admission policies and comparatively high tuition charges which held down their undergraduate enrollments as compared with those of the corresponding public institutions.

Between 1963 and 1970, doctoral-granting institutions experienced a considerably less rapid increase in enrollment (36 percent) than did higher education as a whole (78 percent), and their share of total enrollment fell from 39 to 30 percent. Moreover, enrollment in the institutions with heavy emphasis on research was growing less rapidly than it was at other doctoral-granting institutions.

Comprehensive colleges Our second group of institutions accounted for 31 percent of the students and 16 percent of the institutions of higher education in 1970. These institutions differ from liberal arts colleges in that they tend to offer occupational programs such as engineering, business administration, social work, nursing, and education, along with liberal arts programs. They range from complex institutions with a wide variety of programs, on the one hand, to colleges with a very limited selection of programs, often largely confined to the training of teachers, on the other. Many of them were initially established as two-year public normal schools, were converted to state teachers colleges

---

[2] Data on postbaccalaureate enrollment in 1970 are not available.

in the 1920s or later, and became more comprehensive state colleges after World War II (4). A number of them are often called "regional universities."

In the light of the origins of many of these institutions as public normal schools and teachers colleges, it is not surprising to find that 316, or 68 percent, were publicly controlled in 1970 and that the public colleges enrolled 80 percent of the students in this group of institutions.

Many of these institutions offer graduate work leading to a master's degree. In 1968, 13 percent of the students in the public comprehensive colleges and 15 percent of those in the private institutions were enrolled in graduate programs. A few of these institutions offered doctoral programs, but the number of doctor's degrees awarded by comprehensive colleges in 1967–68 was, in almost all cases, less than 10 each. Many of the states have had formal or informal policies against the development of doctoral programs in state colleges, although in some cases these are being changed. California's policies are particularly restrictive, granting the University of California the sole authority, among public institutions of higher education in the state, to award doctor's degrees, except that it may agree with the state colleges to award joint doctor's degrees in selected fields (5, p. 203).

In contrast with enrollment in the doctoral-granting institutions, enrollment in comprehensive colleges has been rising slightly more rapidly than in higher education as a whole, with the result that the comprehensive colleges' share of the total increased slowly between 1963 and 1970.

Liberal arts colleges Among the four-year institutions of higher education, by far the most numerous are liberal arts colleges. They accounted for 703, or 25 percent, of the institutions of higher education in 1970, but for only 8 percent of the students. It follows, then, that they tended to be small. Their large numbers and small average size reflect the pronounced tendency of Americans, particularly in the nineteenth century, to establish small liberal arts colleges throughout the land.

Most of the liberal arts colleges are private, and the relatively few that are public include some state colleges of recent origin that are likely to grow rapidly and become comprehensive colleges in the near future.

Private liberal arts colleges range from highly prestigious institutions with nationwide reputations to small, struggling, underfinanced colleges that are losing students because of competition with low-tuition public institutions and that may have difficulty surviving in the 1970s if the tuition gap between the public and private institutions continues to widen as it did in the 1960s.

It is also important to recognize that the distinction between many of the liberal arts colleges and comprehensive colleges is not sharp or clear-cut. Liberal arts colleges are frequently extensively involved in training teachers, but typically their students who expect to enter teaching receive a bachelor's degree in a major field in the arts or sciences rather than a bachelor of education degree. Furthermore, as we point out in Appendix A, deciding whether to classify a college as a liberal arts or a comprehensive college was necessarily judgmental in some cases.

Some of the liberal arts colleges have modest master's degree programs, and a few, like Bryn Mawr, award significant numbers of Ph.D.'s. But, for the most part, these are undergraduate institutions, the more prestigious of which number among their graduates large proportions who later receive advanced degrees at leading universities. Only 4 percent of all the students enrolled in liberal arts colleges in 1968 were pursuing graduate studies.

Although liberal arts colleges experienced a 22 percent increase in enrollment between 1963 and 1970, this was far below the 78 percent increase for higher education as a whole in the seven-year period. The share of liberal arts colleges in total enrollment fell from 12 to 8 percent as a result.

Two-year institutions The most rapidly growing institutions in American higher education are the two-year colleges and specialized institutes. They accounted for 1,061, or 38 percent, of all the institutions and for 28 percent of the students in 1970. Policies for their continued growth and development were recommended in the Commissions's report *The Open-Door Colleges: Policies for Community Colleges.*

Among two-year colleges, there were 805 public institutions in 1970. They accounted for 76 percent of the two-year institutions and for 94 percent of the students in two-year colleges. They were typified by the comprehensive public community college, which offered a variety of academic (transfer), general education, and occupational programs. But among the public two-year institutions, there were also about 130 two-year branch campuses of universities and 100 two-year specialized institutes in 1970.[3]

The private two-year colleges are a heterogeneous group, ranging from struggling underfinanced institutions, on the one hand, to expensive two-year colleges for young women, on the other. Most of the private two-year colleges stress academic programs, but some offer occupational programs, and there are a number of private two-year technical institutes, business colleges, and other specialized schools.

During the years 1963 to 1970, the number of two-year institutions increased from 622 to 1,061, or 71 percent, but the expansion was almost entirely among the public colleges, which rose in number from 374 to 805 during this period, whereas the number of private two-year colleges increased only from 248 to 256. There are indications, also, that many private two-year colleges have been experiencing relatively acute financial problems in the last few years. Although their number had reached 296 by 1968, it fell back to 256 by 1970. Part of this decline, however, was the result of 21 private two-year colleges becoming four-year institutions between 1968 and 1970.[4]

Meanwhile, from 1963 to 1970, total enrollment in two-year institutions nearly tripled, reflecting almost entirely the growth of enrollment in public two-year institutions, and the share of two-year colleges in total enrollment rose from 18 to 28 percent.

Whether the community college movement will stimulate an increase in the number of institutions offering only upper-division work for undergraduates and thus specifically designed to admit transfers from two-year colleges is not yet clear, but this may happen. A recent study of upper-division colleges, published in 1970, indicated that only seven of these institutions were in existence when the study was begun, but that, by the time the study

[3] Our total enrollment in two-year institutions, amounting to 2,348,000 in 1970 (Appendix B, Table 4), exceeds the Office of Education figure of 2,210,000 for that year because we have allocated enrollment in two-year branch campuses of universities to two-year institutions, whereas the Office of Education includes enrollment in these two-year branches in the data on their parent institutions.

[4] Only four public institutions experienced a similar change in status between 1968 and 1970.

was completed, at least eight new upper-divison colleges had reached various stages of legislative or state-based approval (6, p. xii).

Specialized institutions Finally, there is a large and heterogeneous group of 436 specialized institutions, accounting for about 15 percent of all institutions of higher education and for 3.4 percent of the students in 1970 (Table 1). However, data on these specialized institutions must be interpreted with caution. Professional schools are included only when they are entirely separate institutions or when they are located on separate campuses of universities and are reported as separate institutions to the Office of Education. Thus, only 43 of the slightly more than 100 medical schools enrolling students in the fall of 1970 are included. A similar situation prevails for other professional fields, including dentistry, other health

TABLE 1
Specialized institutions of higher education, by type and enrollment, United States, fall, 1970

Institutions	Number	Percent	Enrollment Number, in thousands	Percent
Theological seminaries, bible colleges, etc.	211	48.4	51.5	17.9
Medical schools and medical centers	43	9.9	45.5	15.8
Other separate health professional schools	26	6.0	9.6	3.3
Schools of engineering and technology	32	7.3	55.8	19.4
Schools of business and management	28	6.4	45.2	15.7
Schools of art, music, design, etc.	50	11.5	27.2	9.4
Schools of law	14	3.2	9.6	3.3
Schools of education	8	1.8	9.4	3.3
*Other specialized institutions**	24	5.5	34.7	11.9
TOTAL	436	100.0	288.5	100.0

*Other specialized institutions include maritime academies, military and naval academies, and certain specialized schools, such as the SUNY College of Agriculture at Cornell University and the Institute for Advanced Study at Princeton, which do not fall into any of the other categories.

SOURCE: Adapted from the U.S. Office of Education data by the Carnegie Commission staff.

professions, engineering and technology, business administration, and law.[5]

Multicampus institutions During the postwar period, and to some extent earlier, a significant trend in higher education has been the development and expansion of multicampus institutions (8). The prototype of the multicampus institution is the large state university, such as the University of California, the State University of New York, and The Pennsylvania State University, with their many campuses. But there are also many multicampus state college systems, local public community college districts with more than one campus, and a modest number of private institutions of higher education with more than one campus. All told, multicampus systems accounted for about 6 percent of all institutions and about 41 percent of enrollment in higher education in 1968:

	Multicampus institutions as percent of all institutions of higher education, 1968	Enrollment in multicampus systems as percent of total enrollment in higher education, 1968
Public colleges and universities	12.0	52.3
Universities	60.0	77.2
Other four-year institutions	8.5	41.6
Two-year colleges	5.5	25.1
Private colleges and universities	2.0	9.7
Universities	13.8	18.0
Other four-year institutions	1.4	5.7
Two-year colleges	1.5	4.6
Total colleges and universities	5.7	40.5

SOURCE: Estimates developed by Carnegie Commission staff from U.S. Office of Education data.

---

[5] The situation in education is somewhat different. Most teachers colleges today offer liberal arts as well as education programs and have been classified either as comprehensive colleges or liberal arts colleges. Only if they do not have a liberal arts program have we classified them as teachers colleges. Thus, only eight institutions, with 9,400 students, have been so classified. The Office of Education, which classifies institutions predominantly concerned with teacher training as teachers colleges, even if they also offer liberal arts programs, reported 555,000 students enrolled in teachers colleges in 1965 (7, p. 64).

More than three-fourths of the enrollment in public universities was in multicampus systems, while 42 percent of the students enrolled in public four-year institutions and about one-fourth of those enrolled in public two-year colleges were in multicampus institutions in 1968. Since the multicampus institutions tend to be growing relatively rapidly, the trend toward an increasing proportion of total enrollment in these complex systems of higher education is virtually certain to continue.

RACE Racial minority groups—including black Americans, Mexican-Americans, Puerto Ricans, and American Indians—were seriously underrepresented in American higher education in 1970. This was not true, however, of Japanese-Americans and Chinese-Americans, who were well represented in higher education and are not now educationally disadvantaged.

In 1970, there were 522,000 black students enrolled in degree-credit programs in all institutions of higher education (9, p. 13). Of these, 343,000 were 18 to 21 years of age, representing 21 percent of the black population in that age group, as compared with a corresponding proportion of 36 percent for whites. In the 22-to-24 age group, in which students would be likely to be enrolled in graduate work, only 7 percent of the black population was enrolled, as contrasted with 15 percent of the white population.

However, the increase in college enrollment of blacks in recent years has been encouraging. From 1964 to 1970, the number of black students enrolled in institutions of higher education more than doubled, rising from 234,000 to 522,000 (9, p. 13; and 10, p. 18). Part of this increase was attributable to an increase in the black college-age population, but the chief explanation was a significant rise in black enrollment rates. Stimulated at least in part by civil rights legislation, predominantly white institutions of higher education adopted liberalized admissions policies for blacks and, in many cases, actively recruited black students. As a result, whereas slightly more than one-half of all black students were enrolled in predominantly black colleges—the colleges founded for Negroes—in 1964, only 28 percent of all black students were enrolled in these black colleges by 1970 (11, p. 83).

The sharp drop in the black colleges' share of total black enrollment suggests that these colleges may have great difficulty in holding their own in a period of increasing pressure for integration.

Certainly there is clear evidence that they have severe financial problems. But the Commission believes that the black colleges will have a significant role to play for decades to come, and in its report *From Isolation to Mainstream* it has recommended very substantially increased federal aid for these institutions.[6]

Recent data on the enrollment of other minority groups in institutions of higher education are extremely limited, but a special 1969 survey of persons of Spanish origin in the United States indicated that only 15 percent of those aged 25 to 34 had completed one or more years of college (12, p. 20). Educational attainment data collected in the same year showed that 30 percent of whites and 15 percent of blacks aged 25 to 34 had completed one or more years of college (13, pp. 10, 12).

Minority groups have tended to be even less well represented, relatively, at the graduate than at the undergraduate level. Several sample surveys suggested that in 1968 black students constituted, on the average, less than 2 percent of graduate enrollment at leading predominantly white institutions. By the fall of 1970, when nationwide data were compiled for the first time by the U.S. Office of Civil Rights, blacks represented well over 2 percent of the graduate and professional enrollment in many of the leading centers of graduate education—for example, 4.1 percent at Berkeley, 5.5 percent at Harvard, and 5.2 percent at Yale (14). But there is still a long way to go before these percentages will approach the black population aged 22 to 24 as a proportion of the total population in that age group (11 percent).

SOCIO-ECONOMIC STATUS

That college and university students tend to come from middle- and upper-income families is well known. Thus, the comparatively high proportion of low-income families among blacks and certain other minority groups helps to explain their relative underrepresentation in higher education, but it does not provide a complete explanation. Poor preparation in ghetto schools, language barriers in the case of the Puerto Ricans and Mexican-Americans, relatively large high school drop-out rates of students in these groups, and a lack of a tradition in their families of seeking college education are also important factors.

---

[6] The report on the black colleges includes a list of 105 predominantly black institutions, but six of these were not included in the U.S. Office of Education data for 1968.

Among families with dependent members aged 18 to 24, only about a sixth of those with income under $3,000 had one or more dependent members in college in 1969, as compared with nearly two-thirds of those with income of $15,000 or more, and the proportion with dependent members enrolled rose consistently with increasing income (Table 2). Similar relationships prevailed for both black and white families, but at every income level the percentage of families with dependent members enrolled was considerably lower for the black than for the white families.

Youthful members of low-income families clearly encounter financial problems in seeking a college education, but, as suggested above in the case of blacks and some of the other minority groups, the obstacles are not exclusively financial. Project Talent data shed light on the relative roles of socioeconomic status and high school achievement records in explaining college attendance. High achievement records in high school tend to be correlated with family socioeconomic status, and, at all socioeconomic levels, high achievers are considerably more likely to enter college than low achievers. Even so, low socioeconomic status is a barrier to college entrance at all achievement levels. For example, only 61 percent of male graduates of the high school class of 1961 in the highest achievement quartile and the lowest socioeconomic status quartile entered college in the year following graduation, as compared with 92 percent in the highest achievement quartile and the highest socio-

TABLE 2
Percent of primary families with dependent family members 18 to 24 years old with one or more dependent members in college, by family income, United States, October 1969

Income of families with dependent members 18 to 24 years old	All races	White	Negro
All families			
Number (in thousands)	8,772	7,561	1,113
Percent with dependent members in college	41.7	44.5	22.4
Under $3,000	16.5	19.0	11.6
$3,000 to $4,999	23.9	25.2	19.8
$5,000 to $7,499	32.6	34.3	28.9
$7,500 to $9,999	41.8	41.7	
$10,000 to $14,999	49.0	50.1	35.1
$15,000 and over	65.9	66.4	

SOURCE: U.S. Bureau of the Census, "School Enrollment: October 1969," *Current Population Reports,* ser. P-20, no. 206, October 5, 1970, pp. 31–32.

economic status quartile. For female graduates, the difference was considerably more pronounced—only 42 percent of the high achievers in the lowest socioeconomic status quartile, as contrasted with 87 percent of the high achievers in the highest socioeconomic status quartile, entered college (15, pp. 93–94). In other words, because of factors discussed more fully below, low family socioeconomic status tends to be a more significant barrier to college entrance for women than for men.

Surveys of reasons for not attending college also shed light on the relative role of financial barriers. For example, in a 1959 survey of high school seniors' main reasons for possibly not attending college, "not enough money" was reported as the main reason by 32 percent of the men and 29 percent of the women (16, p. 24).

Seniors' main reasons for possibly not attending college

	Men	Women	Total
Poor high school grades, ability	21	10	15
Not enough money	32	29	30
Prefer to work	14	19	17
Prefer marriage	3	19	11
Not interested enough	18	14	16
Other	6	4	5
No response	6	5	6

However, some of those who gave "prefer to work" or "not interested enough" as the main reason may also have been facing some economic problems in contemplating entering college. And "not enough money" may in some cases be a mask for other reasons, because it seems more acceptable than an admission of, say, lack of ability. The fact that the proportion reporting "prefer marriage" was considerably higher for the women than for the men is of special interest in relation to relatively low enrollment rates of women in higher education.

There is some evidence that the proportion of young people who do not attend college because they "cannot afford it" has been declining. For example, a SCOPE study conducted in the mid-1960s indicated that only 17 percent of male and 22 percent of female high school seniors not planning to attend college reported that the greatest obstacle was that it was "too expensive" (17, p. 33).[7]

---

[7] Seniors planning to attend college and those giving no response have been excluded in computing these percentages.

Rising family income levels and increased availability of student aid have played a role in this change, but the most important factor may have been the increasing availability of open-access community colleges.

SEX   Enrollment rates for women in higher education, as a percentage of the relevant age group, are distinctly lower than for men. We have already noted that low socioeconomic status tends to be a more significant barrier for women than for men and that relatively more women indicate that they "prefer marriage."

Among women aged 18 to 19, for example, only 35 percent were enrolled in college in 1970, as compared with 40 percent of the men. Sexual differences in enrollment rates were much larger for those aged 20 to 21 and for those aged 22 to 24. To a considerable extent, these differences in enrollment rates appear to be explained by the tendency of women to marry at relatively early ages. In all these youthful age groups, relatively more women than men were married in 1969, and although both women and men were comparatively unlikely to be enrolled in college if they were married, the proportions of married women who were enrolled were considerably smaller in all three age groups than the corresponding percentages of married men. And because early marriage is especially prevalent among women in low-income groups, it may also be a major explanation of the substantial gap between enrollment rates of male and female high school graduates in low socioeconomic status groups.

This is not to suggest that marriage is the only obstacle to equal participation of women in higher education. There are many other factors involved, to be discussed more fully in future reports of the Carnegie Commission on Higher Education. A study of the relationship between family size and enrollment rates of family members in higher education, for example, has shown that in blue-collar families, but not in white-collar families, college attendance of girls in the family tends to decline precipitously as the number of boys in the family increases. This suggests that males in blue-collar families have first claim on family resources allotted for higher education (18).

Historically, the reverse has been true in black families—i.e., they have been more likely to encourage higher education for daughters than for sons, because there were more opportunities for black female college graduates in such professions as teaching than for male college graduates. There are some indications that this relationship may be changing, but annual data on educational attainment of black men and women from the *Current Population Survey* of the Bureau of the Census have displayed erratic fluctuations from year to year during the 1960s, reflecting the high risk of sampling error in relation to as small a segment of the population, relatively, as college-educated blacks. Similarly, comparative enrollment rates of college-age black men and women have fluctuated in recent years. Probably the safest statement to make, pending the availability of more reliable data from the 1970 Census, is that the percentages of black men and women aged 18 to 21 enrolled in college are very close.

There is evidence of discrimination against women in admission to institutions of higher education, especially at the graduate and professional levels, and there is a good deal of evidence of discrimination against women on college and university faculties. These relationships also will be examined more fully in future Commission reports.

Of the 3 million women enrolled in institutions of higher education in 1968, 178,000, or 6 percent, were studying in the 256 predominantly female institutions. The vast majority of women, clearly, were in coeducational institutions, and a number of factors —including the declining proportion of enrollment in private institutions, the conversion of some predominantly female institutions to coeducational colleges, mergers of women's colleges with men's colleges, and the development of cooperative programs with men's colleges—were contributing to a decline in the proportion enrolled in distinctly female institutions.

Early in 1971, it was reported that the number of exclusively women's colleges had declined by almost 100 in the previous five years, to about 150 (19). However, in some of the women's colleges, there appeared to be growing resistance to shifting to coeducational status.

You can make a tremendous case that women's liberation has caused an increased level of appreciation of women's colleges. At least it has caused women to think about who they are (20).

If there is controversy about the survival of women's colleges, there is perhaps even more controversy over the future of exclusively or predominantly men's colleges. In 1968 there were 207 men's colleges, with a total enrollment of about 146,000. The number is probably considerably lower today. Such well-known male bastions as Yale and Princeton have recently opened their doors to female undergraduates, and increasingly men's colleges are being subjected to pressure by the U.S. Office of Civil Rights to cease discriminating on the basis of sex in their admissions policies.[8]

REGIONAL DIVERSITY

Largely for historical but partly also for economic reasons, there is pronounced diversity in the types of institutions that are predominant in the various regions, as well as marked regional variation in enrollment rates of the college-age population. The highest rates of undergraduate degree-credit enrollment—well above the national average—were found in the Mountain, Pacific, New England, and Central states in 1968 (Chart 6). Enrollment rates in the North Atlantic, North Midwest, and Southwest states were close to the national average, while the Southeast states fell considerably below the national average in this respect.

However, there is a great deal of migration of students both into and out of the various states. In 1968, more than 1 million students attended college away from their states of residence (22). Some of the regions with high undergraduate enrollment rates, notably New England, tended to rank much lower if only students attending college in their states of residence are included, whereas the Pacific and Mountain states continued to rank high on the basis of enrollment of state residents.

Net in-migration and net out-migration of students tended to follow distinctive regional patterns (Appendix B, Table 6). All the New England states, for example, were net importers of students in relation to their enrollment in 1968, except for Connecticut, which was a sizeable net exporter. All the North Atlantic states, on the other hand, were net exporters of students, except for the District of Columbia, which drew large numbers of students from neighboring Maryland and Virginia suburbs.[9]

---

[8] For an interesting presentation of the view that there is a stronger case for continuation of women's, rather than men's, colleges, see reference 21, chapter 7.

[9] New Jersey, a state which has had a long history of relatively meager support of public higher education, and which ranked fiftieth among the states in percent of per capita income expended on higher education in 1967–68 (23, p. 47), was a large net exporter, with net out-migration amounting to 69 percent of enrollment in the state in 1968. New Jersey students were particularly likely to be enrolled in the neighboring states of New York and Pennsylvania (24, pp. 4–5). Early in 1971, the New Jersey State Department of Higher Education announced a master plan for a massive expansion of public higher education in the state that could cost $500 million.

CHART 6 Regional variations in enrollment patterns, United States, 1968

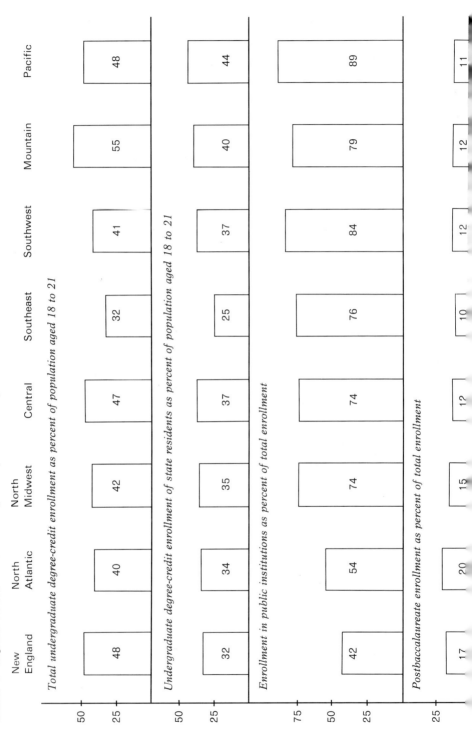

Illinois was the only net exporter of students among the North Midwest states. All the others experienced net in-migration. Most of the Central states, most of the Southeast states, and all the Southwest states were net importers of students. Colorado and Utah stood out as the large net importers in the Mountain states, while, among the Pacific states, California, Oregon, and Washington experienced modest net in-migration, whereas the smaller Pacific states—Alaska, Hawaii, and Nevada—experienced net out-migration.

For the most part, students who migrate enroll in nearby states, but there are some states with outstanding institutions which attract students on a nationwide basis.

Between 1963 and 1968, the proportion of students attending colleges and universities out of their home states declined from 18.2 to 16.8 percent of total degree-credit enrollment (7, p. 66, and 22). It seems probable that a major reason for this decline was the establishment during this period of many new public community colleges, encouraging students to attend college near home, at least for their lower-division work. However, in the last few years there has also been a tendency for some state legislatures to discourage the in-migration of students from other states by establishing quotas and sharply raising tuition charges for out-of-state students at public universities.[10]

To some degree, legislators have tended to blame out-of-state students for instigating campus unrest, but growing financial stringency facing state governments in the last few years has also been a factor in arousing opposition to providing subsidized higher education for out-of-state students.

Interstate migration of students has been an important influence in overcoming the "provincialism" that tended to characterize many state university campuses in the past, and it would be unfortunate if it were seriously impeded.

Regional patterns of variation in types and control of institutions are also of considerable interest. New England, with its long history of development of private colleges, was the only region in which less than half of all students—including undergraduates and graduates

---

[10] Such measures have recently been put into effect or are under consideration in Michigan, Ohio, Texas, and Wisconsin (25). In the spring of 1970, land-grant institutions in the Midwest, unlike those in other regions, reported a 5.3 percent decrease in applications, which was attributed to quotas and tuition increases applicable to out-of-state students (26).

—were enrolled in public institutions in 1968. It was also characterized by a high proportion of students enrolled in doctoral-granting institutions, a relatively small porportion in two-year institutions, and a relatively high percentage of all students enrolled in postbaccalaureate programs.

The North Atlantic states also had a comparatively low proportion—slightly more than half—of their students enrolled in public institutions and a large proportion of all students in postbaccalaureate programs.

In all the other regions, enrollment in public institutions was decidedly predominant—ranging from 74 percent in the North Midwest states to 89 percent in the Pacific states. Other striking characteristics of enrollment were the comparatively large proportions of students enrolled in doctoral-granting institutions in the Southwest and Mountain states, the relatively large proportion in liberal arts colleges in the Central and Southeast states, and the extraordinarily large proportion enrolled in two-year institutions (52 percent, as compared with a national average of 25 percent) in the Pacific states. And, reflecting the historical emphasis on public higher education in the Mountain and Pacific states, enrollment in liberal arts colleges was relatively small in these regions.

METROPOLITAN AND NONMETROPOLITAN AREAS

The probability that a young person would be enrolled in college in 1970 was relatively high if he lived in the outer ring of a metropolitan area and considerably lower if he was a resident of a central city within a metropolitan area or lived outside of a metropolitan area. For example, among young people aged 18 and 19, 42 percent of those living in the outer ring of metropolitan areas, 36 percent of those living in central cities, and 34 percent of those living in nonmetropolitan areas were enrolled. Similar relationships prevailed for other youthful age groups. They also prevailed for the college-age population among both whites and blacks (9, pp. 14–16).

These differences in enrollment rates are largely explained by the fact that family income tends to be higher in the suburbs than in either central cities or nonmetropolitan areas and by the fact that suburban elementary and secondary schools tend to maintain relatively high standards. Young persons in low-income and lower-middle-income families, moreover, are more likely to attend college if there are low-cost public institutions, and especially low-cost, relatively open-access public community colleges, in their communities. It has been found that more than one-half of all high school graduates tend to go on to college if there is a public junior college in the community, whereas one-third or less do so if there is no college in the community (27, p. 27). As we shall find in Section 7, large metropolitan areas vary greatly in the availability of low-cost, open-access institutions, as do nonmetropolitan areas in the various states.

The contrasts in enrollment rates in 1970 between poverty and nonpoverty sections of metropolitan areas with populations of 250,000 or more were, however, much more pronounced than differences in enrollment rates among central cities, suburbs, and nonmetropolitan areas. Among those aged 18 and 19, for example, only 19 percent in poverty areas were enrolled in college, as compared with 44 percent in nonpoverty areas (9, p. 17).[11]

INTERNATIONAL COMPARISONS OF ENROLLMENT RATES

Enrollment rates in higher education in the United States are considerably higher than in any other country. In large part this is explained by the development of mass secondary education in the United States and the expectation that nearly all young people would graduate from high school. Thus, the proportion of the college-age population (76 percent) qualified to enter higher education is considerably higher than in most other countries (Chart 7).[12] Only Canada approaches the record of the United States in this respect, while in Japan about half of the relevant age group completed secondary education in 1965. Where qualified secondary school leavers represent a very small percentage of their age group,

---

[11] The corresponding proportions for whites alone were 21 percent and 45 percent. But among the blacks the difference was not nearly as wide. In the poverty areas, 28 percent of blacks aged 18 to 19, as compared with 29 percent in nonpoverty areas, were enrolled (9, pp. 17–19). In other words, the probabilities of enrollment for both whites and blacks living in poverty areas were low, but a sharp difference by race showed up in the nonpoverty areas.

[12] The fact that the percentage of the population aged 20 to 24 enrolled in higher education exceeds the percentage of the relevant age group completing secondary school in a few countries is probably explained by a rising trend in the number of births in the postwar period, with the result that single-age cohorts in the 20-to-24 age group are smaller than in the age group completing secondary school. It should be noted, also, that the data on enrollment rates in higher education are not precisely comparable, because of the differences in the average age at which students complete secondary school and because of differences in the types of institutions classified as institutions of higher education from country to country. Moreover, the chart applies only to OECD countries, but even if other countries, e.g., the U.S.S.R., were included, the United States would still be far in the lead in enrollment rates.

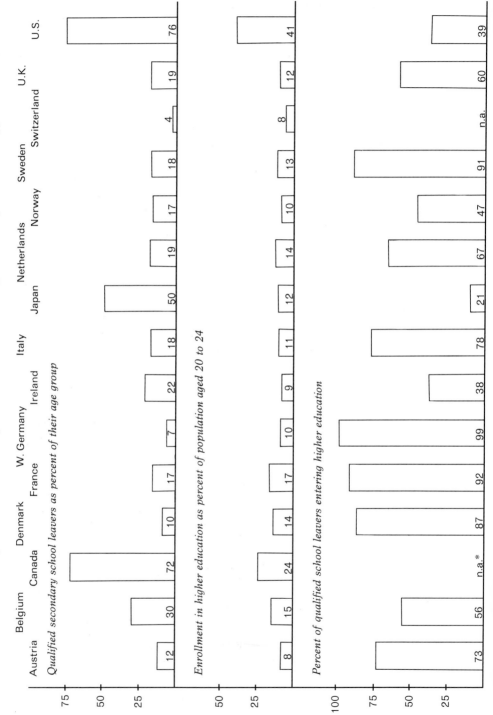

CHART 7 Variations in rates of graduation from secondary school and of enrollment in higher education, OECD countries, years around 1965

as in France and Germany, the porportion of the population aged 20 to 24 enrolled in higher education also tends to be relatively small, but transfer rates to higher education may be very high. In other words, in such countries secondary education may take the form almost exclusively of preparation for higher education.

However, the number *graduating* from college as a percentage of the relevant age group in the United States does not exceed that in other industrial countries by nearly as wide a margin as does the proportion entering higher education, according to data gathered by the Robbins Committee in Great Britain nearly a decade ago, but probably still largely valid as to *comparative* rates today (Table 3).

The leading universities and liberal arts colleges in the United States are highly selective in the admission of students, but higher education as a whole is less selective in varying degrees than the systems of other countries included in Table 3. Thus, by opening wider the doors of our institutions to students of varying degrees

TABLE 3
Percentage of age group entering and completing higher education in selected countries

	Percentage entering, 1958-59* (1)	Percentage completing, 1961-62* (2)	(2)/(1)
Australia	13	9	.69
Canada	24	15	.63
France	9	7	.78
Germany (Fed. Rep.)	6	5	.83
Great Britain	12	10	.83
Netherlands	8	4	.50
New Zealand	15	10	.67
Sweden	11	7	.64
United States	35	17	.49
U.S.S.R.	12	8	.67

*For entering students, the percentage of the age group was computed as a weighted average of the percentages of persons of various ages represented among entrants to higher education, or, where data were not available for this computation, the percentage of the age group from which the largest number of entrants was thought to be drawn was used. For students completing higher education, the percentage of the age group from which the largest number of students completing higher education is drawn was used. For the United States, those completing higher education are persons receiving bachelor's and first professional degrees. Persons receiving the most nearly comparable degrees were included for other countries.

SOURCE: Great Britain, Committee on Higher Education, *Higher Education: Appendix Five to the Report of the Committee Appointed by the Prime Minister under the Chairmanship of Lord Robbins, 1961-63,* Her Majesty's Stationery Office, London, 1964, pp. 9, 11.

of ability, we broaden the opportunity for young people to enter higher education, but we also increase the probability that some of them will drop out or fail to complete the work for the bachelor's degree. In addition, by stimulating the development of occupational programs in two-year colleges, we encourage some students to pursue a program which will terminate at the end of the second year. Other countries likewise in many cases encourage technical education beyond the secondary school level that does not lead to a college or university degree.

The Commission has endorsed an "open access" policy for American higher education in its reports *A Chance to Learn: An Action Agenda for Equal Opportunity in Higher Education* and *The Open-Door Colleges: Policies for Community Colleges.* We believe that there should be open access to the *system* of higher education but not to all institutions. The community colleges should follow an open enrollment policy, admitting all applicants who are high school graduates or who are over 18 years of age and capable of benefiting from continuing education, whereas the admission policies of four-year institutions should be more restrictive. Community colleges should also follow policies of no tuition or very low tuition.

However, certain policies and reforms could be adopted that would increase the probability that a student entering an academic program in either a two-year or a four-year institution would complete the work for a degree. These policies and reforms were discussed in the Commission's report *Less Time, More Options.*

3. The Future of Higher Education: The Questions to Be Examined

The next three decades are likely to be a period of substantial innovation and change in the organization and structure of higher education comparable in significance to two earlier periods of change. The first was the period following the Civil War when many of the leading colleges were transformed into universities. The second was the period since the end of World War II, which was characterized not only by rapid enrollment increases and a steady increase in the share of the public institutions in total enrollment, but also by the emergence of planned state systems of public higher education and of the public two-year community college as the most rapidly growing type of institution.

Along with the continuation of recent trends, we anticipate a new type of development as perhaps the predominant characteristic of the last three decades of the present century—a movement away from participation in formal institutional higher education in the years immediately following high school toward a more free-flowing pattern of participation spread over a broader span of years, perhaps well into middle age and beyond. Students will be encouraged to gain some work experience for several years after high school or after one or two years of college and return to college later, perhaps on a part-time basis, with, it is hoped, more clearly formulated career goals and a better understanding of what additional advanced education might contribute both to the achievement of career goals and to each individual's cultural development. This changing pattern of participation in higher education should, as we recommended in our report *Less Time, More Options,* be encouraged by changes in degree structure; by changes in employer selection policies; and by the development of open universities, external degree systems, and other innovations designed to stimulate a more flexible pattern of higher education experience.

Thus, as we examine in this report the future growth and development of higher education in the United States, we shall be primarily concerned with the manner in which future enrollment increases and needs for institutions might be influenced by such innovative changes. We shall not be especially concerned with the values and functions of higher education, which will be considered in a forthcoming report on academic reform and in the Commission's final report. Nor will we be concerned primarily with problems of financing, which will be considered in future reports on the effective use of resources and on relationships between costs and benefits in higher education. Rather we shall be concerned with such questions as the following:

1 How rapidly will undergraduate and postbaccalaureate enrollments increase if they are influenced primarily by changes in the college-age population and by a continued increase in enrollment rates of the college-age population in line with past trends?

2 How will future increases in enrollment be affected if major policy recommendations of the Carnegie Commission are implemented and if enrollment is subject to other significant influences, such as impending changes in the job market for college graduates and changes in patterns of participation in higher education? Are patterns of enrollment likely to be affected by the cultural revolution that is so greatly affecting the attitudes and styles of life of young people?

3 To what extent should increases in enrollment be absorbed by the growth of existing campuses and to what extent by development of new campuses? Is there an optimum size range for various types of campuses? How can the disadvantages of excessively large size or uneconomically small size be overcome? How can we preserve and encourage diversity in higher education in the face of forces tending toward greater uniformity?

4 To the extent that there will be a need for new institutions, how many institutions of what types will be needed, and where should they be located?

5 In planning for the future, to what extent should we encourage new forms and patterns of higher education?

4. The Future of Higher Education: Assumption A — Largely Uninhibited Growth

ENROLLMENT INCREASES

For the most part, American institutions of higher education managed to expand their resources and facilities to absorb the rapidly increasing numbers of students seeking to enroll in the 1960s. Students who could not qualify for the most selective four-year institutions were admitted to less selective four-year institutions or to two-year colleges. Only toward the end of the decade were there signs of serious stresses and strains resulting from financial stringency in both public and private institutions.

The outlook for smooth absorption of the increased numbers of students who will be seeking higher education in the 1970s is at present very uncertain. Campus unrest, which is leading some state legislatures to "punish" public institutions of higher education by withholding funds and which is causing some alumni and other private donors to hold back on gifts to colleges and universities, may abate somewhat if we withdraw from the Indochina war, but most sophisticated observers do not expect unrest to disappear on campuses. Cutbacks in federal government support of higher education may prove to be temporary if a decline in military expenditures as a proportion of the gross national product facilitates increased appropriations for education and other social services. But a more persistent problem is likely to be the fiscal stringency faced by state and local governments (with the latter representing a significant source of financing of two-year colleges).

Appropriations for higher education must compete at state and local levels with rapidly rising expenditures for welfare, elementary and secondary education, and other public services. State and local governments face serious difficulties in meeting these mounting costs because they tend to rely heavily on sales taxes and, in the case of local governments, property taxes—taxes yielding revenues that tend to rise less rapidly than personal income. In contrast, the tax revenues of the federal government, which rely in large part on personal and corporate income taxes, tend to rise more rapidly than personal income.

In fact, from the perspective of the fall of 1971, it appears likely that higher education will *not* be in a position to absorb the increased numbers of students seeking admission in the 1970s without greatly increased federal government support, along the lines recommended by the Commission in *Quality and Equality: Revised Recommendations, New Levels of Federal Responsibility for Higher Education.* In the absence of such increased federal support, students and their parents in both public and private institutions will have to meet an increased proportion of the rising costs of education through greatly increased tuition and fees. That requirement will be to the detriment of enrollment of many students from low-income families and even of a good many students from middle-income families, and public institutions may continue to be forced to turn away qualified applicants on an increased scale.

Assuming, however, that adequate funds are forthcoming from public sources, that growth is not inhibited by changes in the demand for college graduates or by structural changes in higher education, and that the age distribution of students does not change very much, enrollment trends in the 1970s and the following two decades will be determined by (1) changes in the rate of growth of the college-age population and (2) a continuation of the long-run upward trend in enrollment rates, which in turn reflects primarily the influence of three interrelated and overlapping factors: (a) the upward trend in high school graduation rates, (b) the rise in real per capita income, and (c) changes in the occupational structure which result in an increased demand for persons holding academic degrees.

The Carnegie Commission staff has developed three projections of enrollment, based on U.S. Bureau of the Census population projections[1] and on an extrapolation of trends in enrollment rates for men and women separately from 1947 to 1970 (Appendix B, Table 8). The undergraduate projection is a single series, but we have developed three alternative projections of postbaccalaureate enrollment, based on the following assumptions:[2]

---

[1] Series D.

[2] Three alternative assumptions are also made about the rate of advancement from first-time to intermediate to terminal graduate enrollment. For further details on these projections, see reference 28.

Projection A. That the first-time postbaccalaureate enrollment rate for men, which declined from 1965 to 1969, will return to its 1965 level (58.5 percent of the weighted average number of male recipients of bachelor's degrees in the preceding five years), as draft calls decrease and the number of returning veterans increases, and thereafter will gradually rise at a decreasing rate, reaching 70 percent by the year 2000, and that the first-time enrollment rate for women will rise gradually from its current level of 47 percent to 60 percent by 2000.

Projection B. That first-time postbaccalaureate enrollment rates will rise at one-half the rate assumed under projection A.

Projection C. That first-time postbaccalaureate enrollment rates will remain unchanged from 1969 on.

The projections suggest that the rate of expansion of enrollment in the coming decades will be considerably less pronounced in percentage terms than in the decade of the 1960s. This is likely to be especially true in the 1980s, when total enrollment is projected to decline slightly in the first half of the decade and to increase only about one percent in the second half of the decade (Chart 1).

The reduced rate of growth of enrollment will be attributable primarily to a slowing down in the rate of growth of the college-age population, although our projections also suggest a slight tapering off in the rate of increase in enrollment rates in the 1980s and 1990s (Charts 2 and 3).

During the 1970s those who enter college will largely have been born from 1952–53 to 1962–63. The 1950s were characterized by a relatively high birthrate until 1957, after which a substantial decline occurred (Appendix B, Table 3). Throughout the period from 1952–53 to 1962–63, however, the number of live births was of the order of 4 million a year, or more than yearly births in the preceding decade. Reflecting these relationships, the *number* of students added to enrollment in higher education in the 1970s is likely, actually, to be slightly larger (5.0 million) than the number added in the 1960s (4.7 million) — using our projection of undergraduate enrollment plus postbaccalaureate enrollment projection A. But the *rate* of increase over the decade as a whole (59 percent) will be markedly lower than the exceptionally high rate of increase in the 1960s (124 percent)[3] because the enrollment accretions will be

---

[3] These rates of increase relate to total enrollment, whereas the rates of increase in Appendix B, Tables 1 and 2, relate to degree-credit enrollment.

added to a considerably larger enrollment base. Moreover, the rate of increase in the second half of the decade is projected at only about 17 percent, as compared with 36 percent in the first half, while the total number of students added in the second half (2.0 million) is likely to be significantly lower than the number added in the first half (3.0 million).

The picture in the 1980s will be very different. Those entering college in the 1980s will have largely been born from 1962–63 to 1972–73, a period of generally declining absolute numbers of births as well as of a decline in the birthrate. In the 1990s, on the other hand, we are likely once more to see a sizeable increase in the numbers enrolled in higher education, but the percentage rate of increase for the decade as a whole (about 30 percent) is likely to be far below that of the 1960s and considerably below that of the 1970s.

However, the projections for the 1990s must be regarded as considerably less reliable than those for the 1970s and 1980s. The young people who will enter college from 1970 to 1987 or 1988 have already been born, but our estimates of enrollment for the years after that must be based on projections of the birthrate, which are characterized by a good deal of uncertainty.[4]

We believe that our enrollment projection A, which differs from projections B and C only with respect to postbaccalaureate enrollment, would be the most reliable projection if postbaccalaureate enrollment increases proceeded in accordance with past trends. However, as will be indicated in Section 5, we believe that the changing labor market for Ph.D.'s *and* desirable policy changes affecting postbaccalaureate enrollment will make either projection B or projection C a more likely outcome. In addition, Section 5 will present estimates of the impact of changing labor market conditions and policy recommendations of the Commission on undergraduate enrollment.

ENROLLMENT CHANGES BY TYPE OF INSTITUTION

How will enrollment be distributed among types of institutions in future years? If changes in the 1970s reflect the shifts that occurred from 1963 to 1970, the most rapid growth of enrollment to 1980

---

[4] We have used Census Series D as the most appropriate of the Census Bureau projections of population, largely because it implies slower growth than projections A, B, or C, thus reflecting more recent birthrate trends. But this projection could turn out to be too high if the birthrate continues to decline in the future. Recently the Census Bureau has developed a new set of population projections, Series E, based on lower birthrate projections than those reflected in Series D.

is likely to occur in the two-year institutions. Their enrollment will increase 70 percent and may be expected to increase these institutions' share of total enrollment from 28 to 31 percent (Chart 8 and Appendix B, Table 7). Most of this growth will occur in the public two-year colleges, which are likely to account for 96 percent of all enrollment in two-year institutions in 1980, as compared with 94 percent in 1970.

The comprehensive colleges are also estimated to experience rapid growth. Although their enrollment is likely to increase 58 percent in the 10-year period, however, their share of the total is projected to rise only from 31 to 32 percent. This estimate would be only slightly altered if, as seems likely, some of the public liberal arts colleges were to broaden their programs so that they would be entitled to classification as comprehensive colleges by 1980.

Interestingly, the projections suggest that the most slowly growing group of institutions will be the doctoral-granting institutions, although they will experience a substantial 37 percent increase in enrollment. But their share of the total is likely to fall from 30 to about 27 percent. Moreover, the more prestigious the institution, the less rapid the rate of enrollment growth is likely to be. This reflects the fact that the less prestigious doctoral-granting institutions tend to be younger, and thus in an earlier and more rapid stage of development.

The rate of growth of enrollment in liberal arts colleges is likely to be roughly comparable to that of the doctoral-granting institutions—32 percent. They also will lose out somewhat in terms of their share of total enrollment, and the more selective liberal arts colleges may be expected to grow less rapidly than the other colleges in this category.

The specialized institutions are likely to experience quite substantial enrollment increases, averaging 54 percent by 1980. But their share of total enrollment is likely to remain fairly constant.

Our projections assume that there will be no change in shares of types of institutions in enrollment during the stationary enrollment years of the 1980s, but that from 1990 to 2000 changes in enrollment shares will once again occur at the same rate as from 1963 to 1970. On this basis, enrollment in doctoral-granting institutions is likely to have fallen from three-tenths to about one-fourth of the total by the year 2000, while enrollment in two-year institutions is likely to have risen from 28 to 32 percent. The shares of other types of institutions will probably remain unchanged.

New students and new places 46

CHART 8 Changes in distribution of total enrollment in higher education, by type of institution, actual, 1963 and 1970, and projected, 1980 to 2000

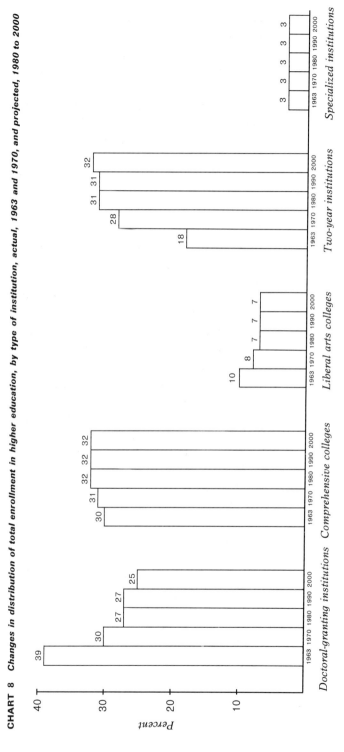

SOURCE: Appendix B, Table 7, and, for 1963, Carnegie Commission staff.

The projections indicate a continued rise in the share of public institutions in total enrollment, from 75 percent in 1970 to 79 percent in 1980 and to 81 percent in the year 2000. In fact, if the stationary trend in enrollment in private institutions that has characterized the last few years were to continue, the rise in the share of public institutions in total enrollment would be even sharper than our projections indicate. The Commission believes, however, that policies should be developed to preserve and strengthen the financial condition of the private institutions, so that they will be able to maintain a reasonably stable share of total enrollment.[5] Yet we must recognize that, even if such policies are adopted, it may be impossible to prevent a slow decline in the share of private institutions as a whole in total enrollment, because the overwhelmingly public two-year college segment of higher education is growing so much more rapidly than the other segments. Even if the private institutions in *each* of our five major categories were to hold their 1970 share of enrollment, overall enrollment in private institutions as a share of the total would nevertheless decline from 25 percent of total enrollment in 1970 to 24 percent in 1980 on the basis of our enrollment projections because of the increasing relative importance of enrollment in two-year colleges. Thus it is probably unrealistic to establish a goal of maintaining the private institutions' share of total enrollment at a constant percentage, but we can set a goal of minimizing the decline in the private institutions' share.

The Commission believes that preserving and strengthening private institutions is of the utmost importance to the future of higher education in the United States. In a number of the private universities and liberal arts colleges, innovative and imaginative approaches to higher education have been developed. The greater freedom of private institutions from political interference helps to preserve academic freedom in the public institutions, and the competition of private institutions helps to improve the quality of education in the public institutions. The Commission is convinced that the quality of education in many of our public institutions of higher education would be much lower than it is today if they were not seeking to model their programs after those in the best of the private institutions and to attract and retain faculty members who would be loath to be associated with public institutions if their quality were distinctly inferior to that of the private institutions.

[5] Such policies have been recommended at the federal level in the Commission's report *Quality and Equality,* and at the state level in its report *The Capitol and the Campus.*

The Commission recommends that the federal and state governments develop and implement policies to preserve and strengthen private institutions of higher education. The federal aid which we have recommended in *Quality and Equality: Revised Recommendations, New Levels of Federal Responsibility for Higher Education* would be available for public and private institutions alike. Policies for state aid to private higher education, with emphasis on student aid as the major approach, were recommended in *The Capitol and the Campus*. We reiterate those recommendations.

5. The Future of Higher Education: Assumption B – Carnegie Commission Goals and Other Influences

CONSTRUCTIVE CHANGE

The Carnegie Commission, as suggested in earlier reports, looks forward to a future for American higher education in which there will be greater equality of opportunity, more options for students, less time devoted to completing the work for degrees, and more opportunities for people to participate in higher education as mature adults. Higher education contributes not only to economic development and to the accumulation of knowledge, but also, in innumerable and often immeasurable ways, to the quality of life in advanced industrial societies.

Among the recommendations made in earlier reports and in the present report, those that would have the most significant effects on future patterns of enrollment and needs for new institutions may be summarized as follows:

1 *Tending to increase enrollment*

- Increased student grants for students from low-income families, cost-of-education supplements to institutions enrolling these students, and a liberalized student loan program *(Quality and Equality)*

- Involvement of institutions of higher education in programs to improve the quality of education in ghetto and rural schools, to bring about more effective recruitment of minority-group students, and to provide additional opportunities for disadvantaged students to overcome any handicaps with which they may have entered during their first two years in college *(A Chance to Learn)*

- Open access to public two-year colleges *(The Open-Door Colleges)*

- 175 to 235 new community colleges by 1980[1]; 80 to 125 of these

[1] In the Commission's report on community colleges, we recommended 230 to 280 new community colleges by 1980, based on analysis of 1968 enrollment

to be in large metropolitan areas (*New Students and New Places,* Section 7)

- 85 to 105 new comprehensive colleges by 1980; 60 to 70 of these to be in large metropolitan areas (*New Students and New Places,* Section 7)

2 *Tending to reduce enrollment*
- A three-year bachelor of arts program for qualified students *(Less Time, More Options)*
- A 1- to 1½-year associate of arts program for qualified students *(Less Time, More Options)*

3 *Tending to strengthen the capacity of private institutions to retain their existing share of total enrollment*
- Federal cost-of-education supplements to institutions, public and private, admitting students holding federal grants *(Quality and Equality)*
- State aid to private institutions *(The Capitol and the Campus)*
- Expanded federal aid to black colleges, one-third of which are public and two-thirds of which are private *(From Isolation to Mainstream)*

4 *Tending to broaden the age distribution of students enrolled in higher education*
- Encouraging stop-outs *(Less Time, More Options)*
- Increased emphasis on adult education, especially in the 1980s, when institutions will be in a particularly favorable position to expand adult programs because of stationary college-age enrollment (*New Students and New Places,* Sections 5 and 8)
- An "education security" program *(Less Time, More Options)*
- External degree systems and open universities (*Less Time, More Options* and *New Students and New Places,* Section 8)

data (29, pp. 63–65), but, between 1968 and 1970, there was a net increase of 79 public two-year institutions of higher education.

We have therefore revised downward our estimate of needs for new two-year institutions, but not by as much as 79, because too few of the added community colleges are in large metropolitan areas and in states with a deficient supply of these institutions to meet our goals of adequate geographic distribution. (See the discussion in Section 7.)

5 *Tending to equalize educational opportunity among the states*
- Expanded federal aid to higher education *(Quality and Equality)*
- Increased state effort in states that have lagged in supporting higher education *(The Capitol and the Campus)*

6 *Tending to reduce inequality in the size of institutions, to achieve more effective use of resources, and to preserve and encourage diversity*
- Optimum size ranges for universities, comprehensive colleges, liberal arts colleges, and community colleges *(New Students and New Places,* Section 6)
- Recommendations relating to cluster colleges and consortia *(New Students and New Places,* Section 6)

7 *Tending to improve the quality of instruction in higher education through increased emphasis on a two-year master of philosophy (M.Phil.) degree and a four-year doctor of arts (D.A.) degree, both designed specifically to prepare candidates for college and university teaching*
- The M.Phil. degree *(Less Time, More Options)*
- The D.A. degree *(Less Time, More Options)*

8 *Tending to reduce the length of time required to obtain a doctor's degree and other advanced degrees*
- Recommendations for a four-year D.A. and for a four-year Ph.D. *(Less Time, More Options)*
- Recommendations for a three-year M.D. and a three-year D.D.S. degree *(Higher Education and the Nation's Health)*

ALTERNATIVE ENROLLMENT ESTIMATES   In this section, we shall attempt to develop alternative estimates of future enrollment that will reflect the influence, not only of implementation of Carnegie Commission recommendations, but also of changes in the labor market for college graduates and holders of advanced degrees, possible shifts in attitudes toward participation in graduate education, and other factors.

Factors tending to increase enrollment   The major influences that would tend to increase enrollment above the future levels discussed in the previous section are the Commission's recommendations (1) for greatly increased student aid, (2) for new comprehensive colleges, and (3) for new community colleges. We have developed an estimate of the combined influence of these three recommendations, which is based on the assumption that, if all three are implemented, financial barriers to enrollment in higher education will be largely removed and enrollment rates in each ability quartile for students in each socioeconomic status quartile will come to resemble rates that have prevailed for corresponding ability levels in the highest socioeconomic status quartile. The method draws on the work of Berls (30, pp. 145–204) and Froomkin (31, pp. 125–137).

The Commission has recommended a substantial increase in federal support for student aid, with federal expenditures for this purpose estimated to rise from $2,614 million in 1971–72 to $4,984 million in 1979–80 in constant dollars (32, p. 33).[2] In comparison, federal obligations for student aid amounted to $941 million in 1966–67 (33, p. 107). Implementation of our recommendations would result in an increase in federal student aid available per full-time equivalent student — based on our enrollment projection A — from about $186 in 1966–67 to about $365 in 1971–72. The maximum grant available for an undergraduate from a low-income family would be $1,000 a year, while the maximum available for a graduate student from a low-income family would be $2,000 a year (plus certain supplements).[3]

We estimate that the combined impact of this increased federal student aid and of our recommendations for new public nondoctoral institutions and community colleges would be to raise total enrollment in 1980 above our projection A by about 600,000 to 900,000 students and FTE enrollment by about 470,000 to 700,000 students. As we shall see, however, this increase may be more than offset by other factors making for a decrease in total enrollment.[4]

---

[2] The amount for 1971–72 is estimated by interpolation from data in the source.

[3] An undergraduate student holding a federal educational opportunity grant and receiving added grants from nonfederal sources would be given a supplementary federal grant in an amount matching the nonfederal grants but not exceeding one-fourth of the original federal grant.

[4] Although developed by similar methods, this estimate of total added students is lower than the estimate of 1 million "extra" students presented in the Commission's first report, *Quality and Equality: New Levels of Federal Responsi-*

Factors tending to reduce undergraduate enrollment We estimate that implementation of our recommendation for a three-year bachelor of arts degree would result in a reduction in the number of undergraduates enrolled in higher education by about 10 to 15 percent by 1980, allowing for a possible increase in retention rates and for the fact that some students entering college with less than adequate preparation would require four years to complete the work for the B.A. degree. This would mean a reduction of about 1.0 to 1.5 million in total undergraduate enrollment, and of about 800,000 to 1.2 million in FTE undergraduate enrollment by 1980. The recommendation for a 1-year to 1½-year associate of arts degree would probably not have an appreciable *additional* impact on enrollment, because many students who could complete the work for an A.A. degree in 1 to 1½ years would probably go on for a B.A. degree.

We do not believe that the less favorable job market for college graduates which is anticipated for the 1970s, as compared with the 1960s, is likely to discourage many young people from entering higher education. Although the enrollment rate in higher education declined somewhat during the Great Depression of the 1930s, it seems probable that the decline was associated with financial stringency in the families of potential college students rather than with poor labor market prospects for college graduates. In fact, in a period of exceptionally heavy unemployment, there was a tendency for some young people to go on to college simply because few attractive jobs were available for high school graduates. Similarly, the fact that enrollment data from the *Current Population Survey* of the U.S. Bureau of the Census appear to show some sensitivity to cyclical upswings and downswings probably reflects the influence of difficulty in financing participation in college rather than the impact of depressed labor market prospects for college graduates.

The prospect of a higher-paying job is by no means the only reason for attending college. It may not even be the most important reason in many cases. The cultural advantages, the opening of new avenues of intellectual interest and appreciation, and the enhanced

---

bility for Higher Education, because our enrollment projections are considerably higher than the Office of Education projections on which our earlier estimate was based and thus incorporate larger increases in enrollment rates than the earlier OE projections. (The OE has since raised its projected enrollment series considerably.) The time period of the two estimates also differs.

social prestige associated with the college experience are likely to continue to stimulate rising enrollment rates even if the income differential associated with college graduation declines. Quite apart from genuine individual aspirations, the social pressures impelling young people to go on to college are strong indeed. As Martin Trow has said:

Social mobility across generations now commonly takes the form of providing one's children with more education than their parents had, and the achievement of near-universal secondary education in America by World War II provided the platform for the development of mass postsecondary education since then (34, p. 3).

Perhaps most important, in a less favorable labor market for college graduates, employers are likely to raise their selection standards and require a bachelor's degree even more widely than is now the case. This would have an unfavorable effect on job opportunities for high school graduates who, in a deteriorating employment market, would be likely to experience prolonged unemployment. On the other hand, college graduates are not likely to be unemployed on any substantial scale but rather will have to accept less attractive jobs than they have been able to get in the past. Thus, ironically, the deteriorating job market for college graduates might well have the effect of impelling more high school graduates to go on to college to escape unemployment. These relationships will be explored more fully in the Commission's forthcoming report *College Graduates and Jobs*.

However, shifts in relative job opportunities among occupations are likely to affect choices of undergraduate majors and graduate fields. There is evidence that students tend to adjust their choices of fields to changes in demand and supply in specific occupations (35). Moreover, because demand is likely to rise especially rapidly in the allied health professions and other technical fields, the increase in the proportion of undergraduates enrolling in occupational programs in two-year colleges that has been going on during the last decade may be accelerated. Another factor that could well induce this type of shift, especially in the case of women, is the deteriorating job market for elementary and secondary school teachers. It takes four to five years to earn a teaching credential, but only two years, for example, to become a nurse.

Enrollment in two-year occupational programs rose from 5.4 to

7.8 percent of total undergraduate enrollment (in all institutions) between 1963 and 1968. Our undergraduate enrollment projection suggests that this proportion may reach 11 percent by 1980. But if labor market changes, as well as establishment of new community colleges, result in acceleration of this trend, as seems likely, the proportion in such programs could be considerably greater than 11 percent by 1980. The percentage of all undergraduates enrolled in occupational programs in 1968 varied widely from state to state and tended to be highest in states with well-developed community college systems, such as Washington, with 18 percent of all undergraduates enrolled in occupational programs, and California, with 21 percent. Most of these students were enrolled in two-year occupational programs.

Other factors, in addition to labor market changes, appear to be accelerating the increase in the public two-year colleges' share of total enrollment. Between 1968 and 1970, enrollment in two-year colleges increased relatively more rapidly than an earlier projection developed by our staff, based on 1963–68 trends, had indicated. All the available information suggests that financial stringency in higher education in the 1968–70 period was a significant factor in inducing this shift. Public universities and state colleges, given legislative appropriations that fell far short of providing adequately for increased enrollment and rising educational costs, were turning away qualified students (36, p. 6). At the annual meeting of the American Association of Junior Colleges early in 1971, there were many reports of students "flocking" to community colleges in response to restricted admissions policies of state colleges and universities and sharply rising tuition at four-year colleges and universities, public and private (37, p. 6). Financial stringency was leading many private institutions to raise tuition more frequently than had been their past practice. These tuition increases, along with the economic recession which began about the end of 1969, enhanced the difficulties faced by students and their families in meeting the rapidly rising costs at private institutions.

By April 1971, six of the eight Ivy League colleges and five of the seven sister colleges were reporting reduced applications for the fall of 1971, as compared with applications for 1970 entrance (38, p. 1). There was also evidence that some of these institutions were experiencing a decline in the relative proportion of their students who came from middle-income families, as compared with students from families with incomes well above average and students from low-income families, who could qualify for student aid based on need.

Whether and to what extent campus unrest has been playing a role in inducing enrollment shifts is somewhat uncertain, although there is scattered evidence that some parents, and perhaps students as well in certain instances, have become disenchanted with the prospect of enrollment on campuses that have experienced disruption and violence.

The question that especially concerns us here is whether the rapidly rising percentage of total enrollment in community colleges, and especially of enrollment in their occupational programs, is likely to result sooner or later in a decline in the proportion of undergraduates who go on to upper-division work. Thus far, there is no evidence of a decline in the undergraduate retention rate as measured by the percentage of first-time undergraduate enrollees (including those enrolled in non-degree-credit programs) who receive bachelor's degrees four years later. In fact, the retention rate increased from 44.2 to 46.2 percent between 1958–62 and 1965–69. But state-to-state variations are very wide, and the pattern of variation from state to state suggests a fairly strong probability that this ratio will decline as more states develop community college systems with comprehensive programs. The retention rate varied inversely with the proportion of undergraduate students who were enrolled in two-year colleges—in fact, the correlation coefficient between 1965 state enrollment in two-year colleges as a percentage of undergraduate enrollment and the 1965–69 retention rate was —.566.[5] California, at one extreme, with 60 percent of undergraduate enrollment in two-year colleges in 1965, had a retention rate of 25.3 percent. Florida, Wyoming, and Washington—states with more than 35 percent of undergraduate enrollment in two-year colleges in 1965—had relatively low retention rates, ranging from 30 to 40 percent. At the other end of the spectrum were seven states—Maine, New Hampshire, Vermont, New Jersey, Pennsylvania, Nebraska, and South Dakota—with retention rates from 60 to 70 percent and very low percentages of undergraduates enrolled in two-year colleges. The highest retention rate (77.3 percent) was in the District of Columbia, with 8 percent of its undergraduate enrollment in two-year colleges.[6]

[5] Nevada, with no two-year college enrollment in 1965, was omitted.

[6] Another factor undoubtedly affecting the retention rate in the District is the fact that its five doctoral-granting institutions are all private. Private colleges

If all the new community colleges recommended by the Commission are developed by 1980, it seems likely that two-year college enrollment may well reach about 40 percent of total undergraduate enrollment—as indicated by the highest of the three projections of two-year college enrollment presented in our report *The Open-Door Colleges*. Other factors that are likely to contribute to this result, as suggested earlier, are the recent acceleration in the rise of the two-year colleges' share of total enrollment and the influence of labor market trends, which are likely to encourage an accelerated rate of increase in enrollment in two-year occupational programs as a percentage of total enrollment.

If two-year college enrollment *should* reach 40 percent of total undergraduate enrollment by 1980, the retention rate nationally might fall to about 36 percent.[7] This would imply a reduction in upper-division enrollment of about 600,000, as compared with our projection A, by 1980. This is quite an "iffy" projection in view of the fact that the retention rate has shown no tendency to decline as yet, but the reason may well be that relatively few states have as yet developed strong occupational programs in community colleges. The projection assumes the development of such programs in all states by 1980.

The impact of increased adult enrollment There are indications that enrollment of adults in higher education has been increasing rapidly in recent years, but it is exceedingly difficult to arrive at accurate estimates on the basis of existing data. To the extent that adult enrollment merely rises in accordance with past trends, the future increase is reflected in our enrollment projections. To the extent that there is an increase in the tendency of young people of college age to stop out for a few years and return to higher education later, the resulting increased enrollment of adults would merely replace a reduction in enrollment of college-age youth.

However, it seems highly likely that the encouragement of adult education which is now being increasingly discussed in higher education circles, and on which we make recommendations in Sec-

---

and universities tend to have more restrictive admissions policies and higher retention rates than public institutions.

[7] This is computed from the regression $y = .560 - .487x$, where $x =$ state two-year college enrollment in 1965 as a percentage of total undergraduate enrollment and $y =$ state bachelor's degrees conferred in 1968–69 as a percentage of first-time undergraduate enrollment in 1965.

tion 8, will result in an acceleration of the rise in adult enrollment rates *over and above* that reflected in our projections. Britain's Open University attracted about 41,000 applicants for 25,000 student places in its first year. With a population about four times the size of Britain's and a much higher ratio of secondary school graduates to population, the United States may expect this type of development to attract much larger numbers, assuming it spreads rapidly beyond the few states already initiating it (Section 8).

Another possible assumption might be that the increase in enrollment of adults in the latter part of the 1970s would be just large enough to offset the effects of the decline in the rate of increase of enrollment of college-age persons that will occur for demographic reasons. On this basis, we may estimate that increased participation of adults would raise total enrollment above our projections for 1980 by about 280,000.

All things considered, we believe that a figure of 250,000 to 350,000 would be a reasonable estimate of the "extra" adults enrolled over and above our projection A by 1980. Since most of these adults would be enrolled on a part-time basis, the added FTE enrollment would be much smaller—probably not more than 80,000 to 130,000.

Factors tending to reduce graduate enrollment The Commission has made several recommendations that would have some effect on graduate enrollment and expects to make additional recommendations in forthcoming reports. Labor market changes, especially the deteriorating market for Ph.D.'s, are also having an impact on patterns of graduate enrollment.

There is growing evidence that the number of applications to various law schools and medical schools has increased sharply in the last year or so, in part because students are shifting away from Ph.D. programs and into these professional schools. Such shifts, of course, will not affect total postbaccalaureate enrollment, but only its composition. However, recent surveys in several institutions have indicated that the proportion of undergraduates planning to go on to graduate work has dropped substantially in the last few years.

More critical, particularly if financial stringency in higher education continues, may be decisions by institutions of higher education to cut back on admissions to graduate education. Several private universities, including Harvard and Stanford, have announced that they will voluntarily curtail enrollments in graduate programs over the next few years, and a commission of the Modern Language Association has called for reduced enrollments in Ph.D. programs in English and the foreign languages (39). These influences will be explored more fully in the Commission's forthcoming report *College Graduates and Jobs*, but we believe that they are likely to depress graduate enrollment by 1980 below our high projection A to the level of either projection B or projection C. This would imply a reduction of about 280,000 to 500,000 as compared with projection A.

The combined impact of all these influences would reduce total enrollment about 680,000 to 1,750,000 (Table 4). On a full-time equivalent basis this would amount to a reduction of about 675,000 to 1,500,000. (The lower end of the range is reduced much less than the upper end when converted to an FTE basis, because the influences that would reduce enrollment are estimated to range more widely than those that would increase enrollment, and increased adult enrollment is assumed to take the form of part-time enrollment in most instances.)

In view of the expectation of an essentially stationary college-age population and enrollment in the 1980s, and the fact that we do

TABLE 4 *Summary of the estimated effects of selected Carnegie Commission recommendations and labor market influences on total enrollment by 1980 (numbers in thousands)*

Recommendation or other influence	Projection A (total enrollment)	Effect on enrollment	Revised projection
Student aid and more widespread availability of community colleges and comprehensive colleges	13,500	+600 to +900	14,100 to 14,400
A three-year B.A.		−1,000 to −1,500	12,000 to 12,500
Reduction in upper-division enrollment because of accelerated shift to two-year colleges		−600	12,900
Increased emphasis on adult education		+250 to +300	13,750 to 13,800
Influences depressing graduate enrollment		−280 to −500	13,000 to 13,220
Net effect		−680 to −1,750	11,750 to 12,820

SOURCE: Estimates developed by Carnegie Commission staff.

not anticipate that the shares of various types of institutions in enrollment will change significantly during that decade, the effects of the influences we have been examining on projected enrollment by 1990 will not be greatly different from their effects in 1980 (Table 5). However, we anticipate that adult participation in higher education will go on increasing in the 1980s and that the reduction in graduate enrollment, as compared with our projection A, will range from 380,000 to 680,000. The net impact of all factors incorporated in the estimate will be a reduction of 410,000 to 1,580,000, or 630,000 to 1,440,000 on an FTE basis. (The lower end of the range of reduction in FTE enrollment exceeds that in total enrollment largely because of our assumption that increased adult enrollment will be chiefly on a part-time basis.)

By the year 2000, the college-age population will have increased considerably over its 1990 level, as will the enrollment rate. With a larger college-age population, we estimate that the effect of the removal of financial barriers and of the more widespread availability of comprehensive colleges and community colleges will be to bring in somewhat more "extra" students than in 1980—from 730,000 to 1,100,000 (Table 6). The precise "mix" of student aid that will be appropriate by the year 2000 will probably be some-

TABLE 5 *Summary of the estimated effects of selected Carnegie Commission recommendations and labor market influences on total enrollment by 1990 (numbers in thousands)*

Recommendation or other influence	Projection A (total enrollment)	Effect on enrollment	Revised projection
Student aid and more widespread availability of community colleges and comprehensive colleges	13,300	+600 to +900	13,900 to 14,200
A three-year B.A.		−930 to −1,400	11,900 to 12,370
Reduction in upper-division enrollment because of accelerated shift to two-year colleges		−600	12,700
Increased emphasis on adult education		+500 to +600	13,800 to 13,900
Influences depressing graduate enrollment		−380 to −680	12,620 to 12,920
Net effect		−410 to −1,580	11,720 to 12,890

SOURCE: Estimates developed by Carnegie Commission staff.

TABLE 6 Summary of the estimated effects of selected Carnegie Commission recommendations and labor market influences on total enrollment by 2000 (numbers in thousands)

Recommendation or other influence	Projection A (total enrollment)	Effect on enrollment	Revised projection
Student aid and more widespread availability of community colleges and comprehensive colleges	17,400	+730 to +1,100	18,130 to 18,500
A three-year B.A.		−1,200 to −1,800	15,290 to 16,000
Reduction in upper-division enrollment because of accelerated shift to two-year colleges		−800	16,600
Increased emphasis on adult education		+1,000	18,400
Influences depressing graduate enrollment		−440 to −800	16,520 to 16,960
Net effect		−340 to −1,670	15,730 to 17,060

SOURCE: Estimates developed by Carnegie Commission staff.

what different from the package recommended by the Commission for the 1970s. We believe that, as time goes on, students will become more accustomed to the idea of borrowing to finance their higher education and that student grants will play a relatively reduced role and loans a relatively increased role in the financing of a student's college education. These issues will be discussed more fully in our forthcoming report *Higher Education: Who Benefits and Who Pays?*

With a larger projected undergraduate student body, the reduction in enrollment associated with a three-year B.A. will be greater than in 1980, as will the reduction in upper-division enrollment associated with the increased share of community colleges and their occupational programs in total enrollment. Participation of adults in higher education, over and above the enrollment of adults embodied in projection A, will probably be higher than in 1980 or 1990, as will the reduction in graduate enrollment. All in all, the net reduction in total enrollment is estimated to range from 340,000 to 1,670,000 — or from 730,000 to 1,780,000 on a full-time equivalent basis. The reduction in enrollment is larger on an FTE basis because of our assumption that most enrollment of adults will be on a part-time basis.

COST ESTIMATES

The combined impact of the Carnegie Commission recommendations and other influences, as we have seen, would be to reduce enrollment on an FTE basis to about 675,000 to 1,500,000 less than the level indicated by projection A by 1980. The midpoint of this range is about 1,090,000. On the assumption that this reduction is distributed evenly over the decade—a reasonable assumption—this means that FTE enrollment may be expected to increase by about 109,000 less annually than projection A indicates. Major savings in construction costs of institutions of higher education would result. Assuming that construction costs are likely to increase at an annual average rate of about 5 percent in the 1970s, we may estimate total construction costs of institutions of higher education over the ten years, 1971–72 to 1980–81, as follows:

	Total construction costs, 1971–72 to 1980–81
1. FTE enrollment increases as indicated by projection A	$18.0 billion
2. FTE enrollment increases by 109,000 less per year than indicated by projection A	$12.4 billion

Thus the combined impact of the influences we have been examining would be to reduce total construction costs by about $5.6 billion over the decade. If the Commission's recommendation, in *Quality and Equality: Revised Recommendations, New Levels of Federal Responsibility for Higher Education,* for federal construction grants amounting to one-third of construction costs is implemented, one-third of this saving, or $1.9 billion, would accrue to the federal government and two-thirds, or $3.7 billion, to other levels of government and to private sources of financing.

The problem of estimating the impact of these influences on current educational costs of institutions of higher education is somewhat more complex, partly because graduate education is considerably more costly than undergraduate education. On the assumptions (1) that the cost of education per student at the graduate level is three times that at the undergraduate level and (2) that average educational costs per FTE student will rise at an annual average rate of 5.2 percent a year in the 1970s, we have developed three alternative estimates of the total educational costs of institutions of higher education associated with enrollment increases in 1971–72 and 1980–81.[8] They use, alternatively, projections A, B, and C for graduate enrollment, while the second and third estimates assume that all other FTE enrollment increases by 78,000 less per year than projection A indicates, i.e., is 780,000 less by 1980:

	Total educational costs associated with increased enrollment as compared with the preceding year	
	1971–72	1980–81
1. FTE enrollment increases as indicated by projection A	$ 1.2 billion	$ 1.0 billion
2. FTE graduate enrollment increases as indicated by projection B—all other FTE enrollment increases by 78,000 less per year than indicated by projection A	$960.0 million	$670.0 million
3. FTE graduate enrollment increases as indicated by projection C—all other FTE enrollment increases by 78,000 less per year than indicated by projection A	$880.0 million	$570.0 million

Thus, by 1980–81, the second assumption would yield savings of $330 million per year in current educational costs, while the third assumption would yield savings of $430 million per year, as compared with the first assumption.

The Commission's recommendation that small institutions should be encouraged to grow to at least the minimum enrollments recommended in Section 6 would also yield some savings in educational costs, but these are tricky to estimate because many of the public institutions will grow without any special encouragement in the form of differential state or federal aid. Thus an estimate of these savings would have to distinguish between institutions that are likely to grow to our proposed minimum enrollment levels

---

[8] Educational costs include expenditures for instruction and departmental research, extension and public service, libraries, plant maintenance and operation, general administration, organized activities of educational departments, other sponsored programs, and other educational and general expenses. These costs per FTE student rose at an annual average rate of about 5.4 percent in the 1960s, but we are assuming a slightly reduced rate of increase for the 1970s on the ground that, in a deteriorating market for college faculty, faculty salaries are likely to rise less rapidly than in the 1960s.

in any event and those that would need special aid to achieve this objective or would be able to achieve it only by merging with another institution. We plan to present estimates of these savings in our forthcoming report *Effective Use of Resources in Higher Education.*

6. The Growth of Institutions

OPTIMUM SIZE

American institutions of higher education range in size all the way from colleges with less than 100 students to gigantic campuses of major universities with 40,000 or 50,000 students.

Chart 9, based on total enrollment, and Table 7, based on full-time equivalent enrollment, indicate that public doctoral-granting institutions tend to be decidedly larger than their private counterparts. None of the 101 public doctoral-granting institutions had less than 3,000 FTE students in 1970. About one-half had 15,000 or more, while about 23 percent had 20,000 or more. Their median FTE enrollment was about 15,000. Only about 10 percent of the private doctoral-granting institutions had 15,000 or more FTE students, and the median FTE enrollment of the private institutions was about 7,000.

The public comprehensive colleges ranged widely in size—from less than 1,000 FTE students to 20,000 or more in a few cases—but the majority ranged from 2,500 to 10,000, and their median FTE enrollment was about 4,300. The private comprehensive colleges tended to be somewhat smaller, with a median FTE enrollment of about 2,300 and with the great majority ranging in number of FTE students from 1,500 to 5,000.

Liberal arts colleges, which, as we have seen, were established all over the country during the course of the nineteenth and early twentieth centuries, tend to be relatively small. About two-thirds of the private liberal arts colleges had less than 1,000 FTE students in 1970, and their median FTE enrollment was about 800. The public liberal arts colleges tended to be slightly smaller, but it will be recalled that, as public colleges increase in size, they tend to add professional programs and to be classified as comprehensive colleges.

The widest range in size is found among public two-year colleges.

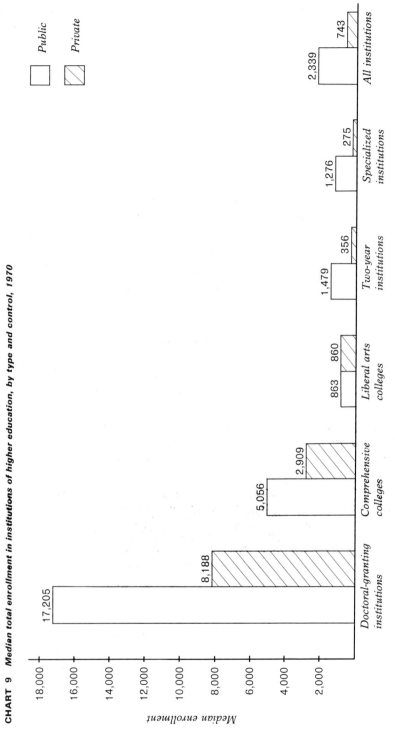

CHART 9 Median total enrollment in institutions of higher education, by type and control, 1970

SOURCE: Carnegie Commission staff.

In 1970, about 20 percent of these institutions had less than 500 FTE students, while, at the other end of the spectrum, about 8 percent had 5,000 or more. But most ranged from 500 to 5,000, and their median FTE enrollment was about 1,100. Many of the small public community colleges were newly established and could be expected to grow rapidly, but some tended to be small because of their location in small communities.

Far smaller were the private two-year colleges, with their median FTE enrollment of about 330. Approximately 73 percent of these colleges had FTE enrollments of less than 500.

The great majority of the specialized institutions were private, and these tended to be small. About three-fourths had less than 500 students, and their median FTE enrollment was only about 230.

To round out the picture, public institutions of higher education in the United States tend to be considerably larger than private colleges and universities, with an overall median FTE enrollment of about 1,800, as compared with about 660 for the private institutions, in 1970. Of considerable interest, also, is the fact that between 1968 and 1970, the median size of nearly all our groups of public institutions increased, whereas the median size of most of the private groups declined slightly.

The Commission believes, with some qualifications, that there is an optimum size range for each major type of institution of higher education. Colleges and universities which are too small cannot operate economically, while, beyond a given size, there may be minimal additional economies of scale, and the institution may become too large to provide an intellectually challenging environment for many students.

For some time, the Commission has been conducting studies of economies of scale in higher education. The results of these studies will be presented briefly here, and more fully in the Commission's forthcoming report *Effective Use of Resources in Higher Education.* Charts 10 through 16 show educational expense per FTE student on the vertical axis and FTE enrollment on the horizontal axis. We have included educational and general expense less expenditures for organized research in our measure of educational costs—i.e., (1) instruction and departmental research, (2) extension and public service, (3) libraries, (4) plant maintenance and operation, (5) general administration, (6) organized activities of educational departments, (7) other sponsored programs, and

TABLE 7 Institutions of higher education, by type, control, and FTE enrollment, fall 1970

Enrollment	Doctoral-granting institutions		Comprehensive colleges		Liberal arts colleges	
	Public	Private	Public	Private	Public	Private
Total						
Number	101	63	316	147	27	676
Percent	100.0	100.0	100.0	100.0	100.0	100.0
0–249	—	1.6	—	—	7.4	8.1
250–499	—	—	—	—	11.1	14.8
500–999	—	1.6	0.6	2.0	55.6	44.7
1,000–1,499	—	3.2	8.9	13.6	—	21.4
1,500–1,999	—	1.6	8.2	25.2	—	7.4
2,000–2,499	—	1.6	9.2	16.3	18.5	2.7
2,500–2,999	—	3.2	6.6	14.3	3.7	0.6
3,000–4,999	3.0	19.0	24.7	18.4	3.7	0.3
5,000–7,499	13.9	22.2	18.7	7.5	—	—
7,500–9,999	11.9	20.6	10.7	2.7	—	—
10,000–14,999	21.8	15.9	8.2	—	—	—
15,000–19,999	26.7	3.2	3.2	—	—	—
20,000–29,999	14.8	6.3	1.0	—	—	—
30,000 or more	7.9	—	—	—	—	—
Median FTE enrollment	14,885	7,053	4,340	2,282	784	804

SOURCE: Data adapted by Carnegie Commission staff from U.S. Office of Education data.

(8) other educational and general expense. We have also developed data relating to instruction and departmental expenditures alone per FTE student, but we have found the broader measure of educational expense presented in our charts to be a superior measure for purposes of analysis of economies of scale, because there are significant economies in such items as library expense and plant maintenance.

Doctoral-granting institutions Educational expenditures per full-time student tend to fall quite sharply until FTE enrollment reaches about 5,000 and more gradually to increasing size beyond that point (Chart 10).[1] However, it is important to recognize that

---

[1] A similar pattern of cost behavior has been revealed in studies of costs of certain undergraduate programs in doctoral-granting institutions in the Province of Ontario (40).

Two-year institutions		Specialized institutions		Total	
Public	Private	Public	Private	Public	Private
805	256	64	372	1,313	1,514
100.0	100.0	100.0	100.0	100.0	100.0
5.7	40.2	20.3	54.6	4.6	23.9
13.8	32.4	7.8	21.0	9.1	17.2
26.1	18.4	31.3	15.6	18.8	27.1
15.8	5.1	12.5	3.7	12.4	12.8
8.8	1.9	10.9	2.4	7.9	6.7
7.2	—	—	1.1	7.0	3.1
5.0	1.2	4.7	0.3	5.0	2.1
10.0	0.8	9.4	1.3	12.9	3.2
4.0	—	3.1	—	8.2	1.7
2.6	—	—	—	5.1	1.1
0.9	—	—	—	4.2	0.7
—	—	—	—	2.8	0.1
0.1	—	—	—	1.4	0.3
—	—	—	—	0.6	—
1,140	326	857	229	1,820	664

we are describing only a general tendency. The scatter about the fitted curve in Chart 10 is very wide.

Our data shed a good deal of light on the reasons for this pattern of behavior of costs. Faculty-student ratios tend to decline, and there are declining administrative, plant maintenance, and library costs per FTE student up to about 15,000 FTE enrollment. Beyond that point, there is some tendency for these measures to rise with increasing enrollment, but there are too few institutions represented in the larger-size groups to draw firm inferences on this point.[2] However, throughout the size range of doctoral-granting institutions, there is a tendency for both the ratio of graduate enrollment and the number of fields in which degrees are awarded to rise with

[2] For purposes of this analysis, we were not able to include some of the large multicampus institutions, which reported financial data only for the entire institution, not by campuses.

CHART 10 *Educational expenditures* of doctoral-granting institutions per full-time equivalent student, by enrollment, 1967–68*

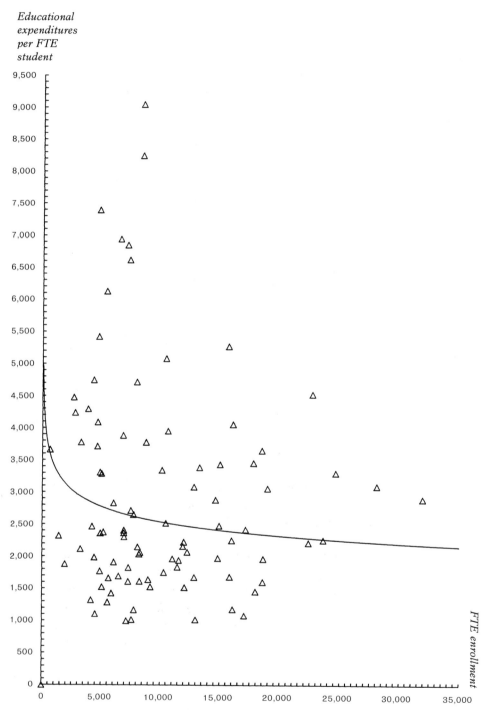

*Includes expenditures for instruction and departmental research, extension and public service, libraries, plant maintenance and operation, general administration, organized activities of educational departments, other sponsored programs, and other educational and general expense. A few institutions with very high expenditures per student have been omitted.

SOURCE: Carnegie Commission staff.

increasing enrollment. In addition, and clearly associated with the rise in the proportion of graduate students, there is a tendency for the number of faculty members per field to rise with increasing size. These factors make for rising costs per student as the size of the institution increases and offset, at least in part, the factors making for economies of scale. At a given point, the importance of the factors making for rising costs appears to outweigh the significance of the factors making for decreasing costs.

Despite some evidence of economies of scale within a given size range, the data must be interpreted with caution, particularly with respect to private doctoral-granting institutions I and II—the research-oriented private institutions (not shown separately). Costs per FTE student are considerably higher for this group than for any of our other groups of doctoral-granting institutions, although the difference tends to narrow with increasing size. Also of significance is the fact that some of the variables discussed above do not display the same pattern of variation with increasing size as in the case of the other groups of doctoral-granting institutions. The ratio of graduate enrollment to total enrollment is relatively high in all size classes and shows no consistent pattern of change with increasing size. Similarly, faculty-student ratios and the number of faculty members per field display no consistent pattern of variation with increasing size. The explanation appears to be that many of the relatively small institutions in this group tend to have high costs of education per student, large numbers of faculty members per field, and high ratios of graduate students, because they have large incomes—they are well endowed, receive sizeable gifts, and have ready access to research funds. Thus they do not need to grow larger to take advantage of economies of scale—they have enough income to meet their relatively high costs. At least, this has been the pattern over much of their history and still tended to be the case in 1967–68, the year on which our analysis is based. Currently, a number of these wealthy institutions are facing serious financial stringency. Yet there is little evidence that they are responding by increasing enrollment. In fact, they are tending to respond by various belt-tightening measures and in some cases by cutting back on graduate programs. The point to be stressed in this context is that wealthy small institutions are not necessarily operating uneconomically simply because of small size. On the other hand, small private institutions that are poorly endowed and heavily dependent on tuition are in a very difficult situation, because they cannot charge tuition that is high

CHART 11 *Educational expenditures* of comprehensive colleges I per full-time equivalent student, by enrollment, 1967–68*

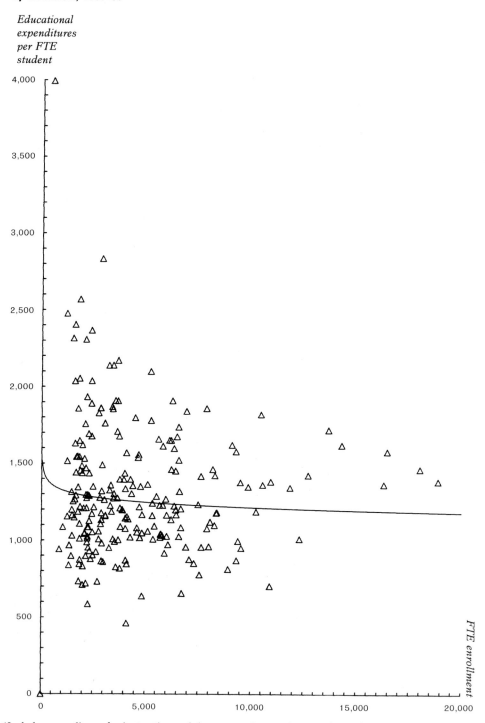

*Includes expenditures for instruction and departmental research, extension and public service, libraries, plant maintenance and operation, general administration, organized activities of educational departments, other sponsored programs, and other educational and general expense.

SOURCE: Carnegie Commission staff.

enough to cover their high costs without pricing themselves out of the market.

Comprehensive colleges The more comprehensive of these institutions (comprehensive colleges I) display a very slight tendency toward declining educational costs per FTE student to about 5,000 FTE enrollment (Chart 11). Beyond that size, costs almost level off. Again, however, as in the case of the doctoral-granting institutions, there are wide deviations from our fitted curve. (It should be noted—Appendix A—that no institutions with less than 2,000 students in 1970 were classified in this group.)

More detailed data shed considerable light on why costs behave in this manner. The reasons are somewhat similar to those that prevail in the case of doctoral-granting institutions, but the disappearance of economies of scale occurs with lower enrollment. All components of educational expense per FTE student, as well as faculty-student ratios, either level off or continue to decline when FTE enrollment increases beyond about 8,000. However, the number of fields in which degrees are offered tends to increase steadily with rising enrollment, and the larger colleges are likely to offer relatively expensive programs, such as engineering and allied health programs. This propensity to add relatively costly programs as enrollment increases apparently tends in large part to offset the effects of declining faculty-student ratios, declining administrative costs per student, and other factors contributing to economies of scale. In fact, instructional costs per FTE student, apparently for this reason, show no economies of scale beyond about 2,000 to 2,500 students. Economies of scale in this group of institutions occur almost entirely in the noninstructional components of cost, such as administrative expense.

For comprehensive colleges II, the group with more limited programs, costs tend to be somewhat lower than in the more comprehensive institutions and to decline quite sharply with increasing enrollment to about 2,500 FTE enrollment (Chart 12). Beyond that point they tend to decline more gently. However, it is important to keep in mind the fact that smaller institutions, i.e., institutions with 1,000 to 2,000 students, are represented in this group (Appendix A) than in comprehensive colleges I. Hence, the enrollment range within which costs can fall sharply is represented in comprehensive colleges II, but not in comprehensive colleges I. By definition, also, comprehensive colleges II do not

CHART 12 Educational expenditures* of comprehensive colleges II per full-time equivalent student, by enrollment, 1967–68

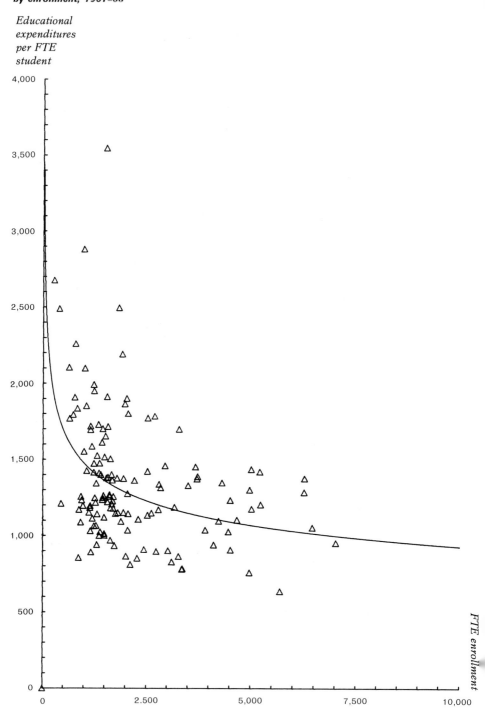

*Includes expenditures for instruction and departmental research, extension and public service, libraries, plant maintenance and operation, general administration, organized activities of educational departments, other sponsored programs, and other educational and general expense.

SOURCE: Carnegie Commission staff.

add more and more relatively expensive programs as they grow larger.

Liberal arts colleges Liberal arts colleges I tend to have considerably higher costs per student than liberal arts colleges II (Charts 13 and 14). Educational costs per FTE student in the first group of liberal arts colleges tend to decline sharply to about 900 FTE enrollment and much more gently to about 2,000 FTE enrollment. Very few institutions with more than 2,000 FTE enrollment are represented in this group. There are fairly steady decreases in all components of costs, as well as in faculty-student ratios, with increasing size.

In the case of this group of institutions—all private—as in the case of private research universities, many of the smaller colleges can afford high educational costs per student because they are well endowed and have generous alumni. Thus they can afford to remain small by choice. Even so, they may not be developing the same educational product as larger institutions. Their faculty may be too small to offer their students as wide a range of courses as is available in somewhat larger schools. However, some small liberal arts colleges, as we shall see, overcome this difficulty through consortium arrangements with neighboring institutions.

In our second group of liberal arts colleges, the evidence of economies of scale is very mixed. Among the small colleges in this group, there is an exceedingly wide range in costs. Some are relatively well endowed and can afford to spend relatively large amounts per student, whereas many others are in a tight financial situation and have comparatively small expenditures per student. Interestingly, our more detailed data show that any economies of scale in the private institutions in this group are almost entirely in noninstructional costs. Faculty-student ratios decline with increasing size, but the number of fields in which degrees are offered rises sharply with increasing size, the number of faculty members per field rises moderately, and average faculty salaries rise consistently with increasing size. Thus factors making for economies of scale in instructional costs tend to be offset by factors making for increased costs per student with increasing size. Many of the smaller institutions in this group are severely hampered in the choice of fields that can be offered to their students.

CHART 13 *Educational expenditures* of liberal arts colleges I per full-time equivalent student, by enrollment, 1967–68*

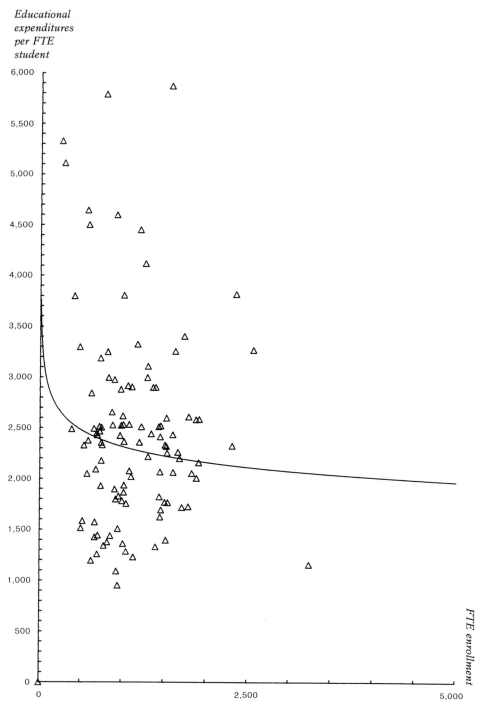

*Includes expenditures for instruction and departmental research, extension and public service, libraries, plant maintenance and operation, general administration, organized activities of educational departments, other sponsored programs, and other educational and general expense.

SOURCE: Carnegie Commission staff.

CHART 14 *Educational expenditures* of liberal arts colleges II per full-time equivalent student, by enrollment, 1967–68*

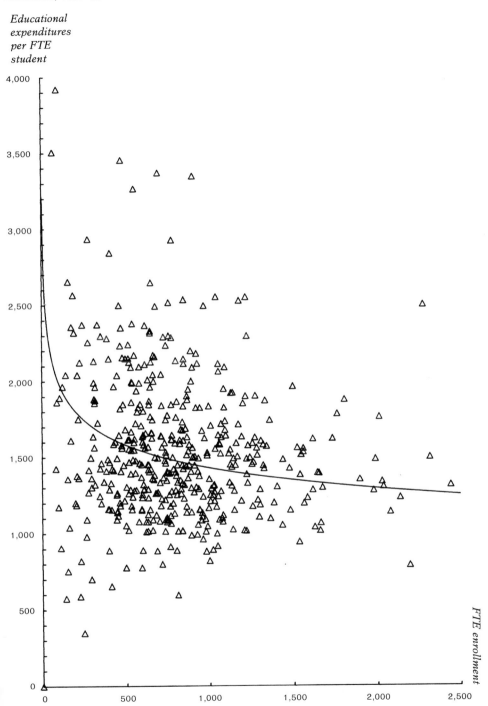

*Includes expenditures for instruction and departmental research, extension and public service, libraries, plant maintenance and operation, general administration, organized activities of educational departments, other sponsored programs, and other educational and general expense. A few institutions with very high expenditures per student have been omitted.

SOURCE: Carnegie Commission staff.

Two-year institutions Public two-year colleges display some economies of scale to about 2,000 FTE enrollment, but their educational costs per student tend to stabilize beyond that point (Chart 15). Again, there is a very wide scatter around the fitted curve. The more detailed data for public two-year colleges indicate many of the same relationships found in the case of comprehensive colleges I. Faculty-student ratios decline with increasing size, as do the various noninstructional components of cost, but the number of fields in which degrees are offered increases, as does the number of faculty members per field. And, with increasing enrollment, there is a tendency, as in the case of comprehensive colleges I, to add relatively expensive programs such as engineering technology and allied health training. However, the data for public comprehensive colleges I show only a slight tendency for average faculty salaries to rise with increasing size, whereas the rise is quite pronounced in community colleges. This reflects the fact that there is a decided tendency for community colleges with large enrollments to be located in large cities, where high salaries must be paid to attract and retain faculty members and nonacademic personnel. Small community colleges, on the other hand, tend to be located in small communities where salary and other costs tend to be lower.

Small private two-year colleges have higher average costs than their public counterparts, but their costs approach those of the public colleges as they increase in size to about 1,000 FTE enrollment (Chart 16). As we pointed out in Section 2, the average size of these institutions is very small. Thus the fact that average educational costs per FTE student were considerably higher than for the public two-year institutions was explained primarily by their smaller average size.

There is a good deal of evidence that private two-year colleges, as a group, are very "squeezed" financially, though some of them attract students from high-income families and levy very high tuition charges. At the other end of the spectrum, however, are the great majority of private two-year institutions. Average faculty salaries in private two-year colleges were about $2,000 below those in public two-year colleges in 1967–68, and, as in their public counterparts, average faculty salaries tended to rise with increasing size.

The results of our economies of scale analysis suggest that there is a minimum size below which each type of institution cannot

CHART 15 *Educational expenditures* of public two-year institutions per full-time equivalent student, by enrollment, 1967–68*

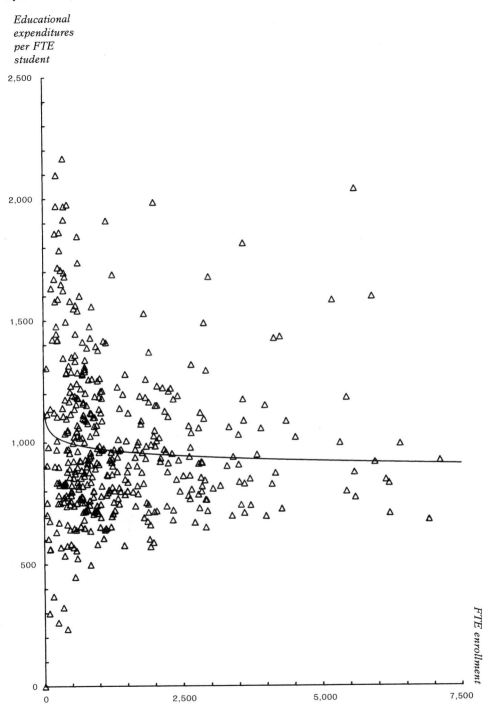

*Includes expenditures for instruction and departmental research, extension and public service, libraries, plant maintenance and operation, general administration, organized activities of educational departments, other sponsored programs, and other educational and general expense. A few institutions with very high expenditures per student have been omitted.

SOURCE: Carnegie Commission staff.

CHART 16 Educational expenditures* of private two-year institutions per full-time equivalent student, by enrollment, 1967–68

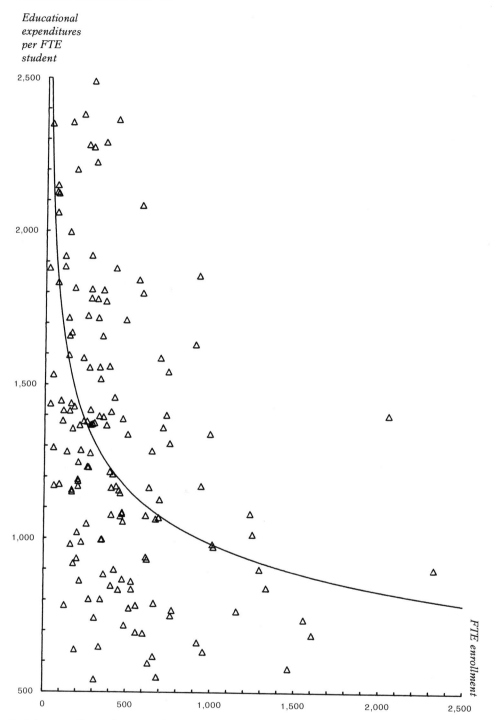

*Includes expenditures for instruction and departmental research, extension and public service, libraries, plant maintenance and operation, general administration, organized activities of educational departments, other sponsored programs, and other educational and general expense.

SOURCE: Carnegie Commission staff.

operate economically, unless it is exceptionally well endowed, and even then it may not be in a position to offer its students as broad a range of courses as somewhat larger institutions are in a position to provide.

There are, however, disadvantages in allowing institutions to become too large. In a very large institution, it becomes extremely difficult for undergraduate students, and even in some cases for graduate students, to have contacts with faculty members outside the classroom. The average student has fewer opportunities for leadership in student activities than on smaller campuses. Faculty members tend to have contacts primarily with members of their own departments and to some extent with members of closely related departments, e.g., physicists with other physical scientists. In short, neither students nor faculty have much of a sense of belonging to an academic community. Perhaps this is one of the reasons why disruption has been found to be more prevalent in recent years on very large campuses than on smaller campuses (41, pp. 64-65). Furthermore, a survey of what happened on campuses following the Cambodian invasion of May 1970 showed that the proportion of campuses reporting that the invasion had a "significant impact" on campus operations varied directly with size of the campus—from 41 percent of those with less than 1,000 students to 90 percent of those with more than 12,000 students (42).

Very large campuses can also be extremely difficult to administer. If relatively uniform standards are to apply, for example, to such matters as the selection and promotion of faculty members, inordinate amounts of time are required on the part of faculty and administrators to ensure that these standards are applied equitably in all departments. If the administration needs to consult with representatives of the faculty or students, especially in times of crisis, it can be exceedingly difficult to determine how to ensure that those consulted are truly representative of diverse faculty and student opinion. More generally, the flexibility and adaptability that are desirable in an intellectual environment become increasingly difficult to achieve as the faculty and student body become larger and larger.

Beyond a certain size, moreover, additional enrollment increases are not likely to contribute to increased quality.[3] Enrollment in

---

[3] For example, many of the universities rated as having very high quality graduate programs in the recent American Council on Education study (43) are relatively small, while many of the liberal arts colleges in our selectivity I group are quite small.

each of our major types of institutions should in most cases grow to a point at which it is economically feasible to maintain a faculty of the size and breadth required for the programs the institutions desire to offer, but enrollment increases beyond that point may increase diversity in faculty specialties without adding appreciably to the quality of education provided students.

In addition, the Commission believes that a state plan for higher education will provide more adequate geographic distribution of institutions if it sets a reasonable maximum on the size of existing campuses of various types and plans on new campuses to absorb enrollment increases beyond that point.

An optimum size range for each type of institution cannot be determined on the basis of economies of scale analysis alone. For example, in the case of comprehensive colleges, there appear to be only minor *net* economies of scale beyond about 2,000 to 2,500 FTE enrollment, but a college needs to grow to about 5,000 on an FTE basis or to about 6,400 on a total enrollment basis to offer a truly comprehensive program. All things considered, we suggest that colleges will run a risk of failing to take advantage of economies of scale and/or of not offering their students an adequate choice of programs if they do not reach minimum enrollments approximately as follows:

	FTE enrollment	Total enrollment [4]
Doctoral-granting institutions	5,000	5,900
Comprehensive colleges	5,000	6,000
Liberal arts colleges	1,000	1,100
Two-year institutions	2,000	2,500

Although our economies-of-scale analysis is helpful in suggesting minimum enrollments, it does not provide a basis for determining appropriate maximum enrollments, because a campus may become too large to provide an intellectually challenging environment for its students before it reaches the point of diminishing economic returns to additional enrollment. For this reason, any suggestions the Commission makes on maximum size must be largely judgmental and somewhat tentative. It should also be recognized that

---

[4] For each type of institution, the relationship of FTE enrollment to total enrollment is determined on the basis of the average ratio of part-time to total enrollment in that type of institution.

there are disadvantages in an excessively rapid rate of growth, just as there are some disadvantages in absence of growth. It is difficult for an institution to adjust to a rate of growth that is excessive, whereas it is easier to achieve innovations in a period of some growth than in a period of stationary or declining enrollment.[5]

The Commission suggests the following maximum enrollments, recognizing that special considerations in individual situations may be a basis for modification, and that many private institutions are not likely to reach these maxima:

	FTE enrollment	Total enrollment
Doctoral-granting institutions	20,000	23,500
Comprehensive colleges	10,000	12,000
Liberal arts colleges	2,500	2,700
Two-year institutions	5,000	6,200

Public institutions, with the exception of those located in very small communities, are not likely to encounter serious difficulties in reaching the suggested minimum enrollments as the college-age population continues to increase in the 1970s. For the most part, small public institutions are in an early stage of their development. The suggested *maxima* will pose more serious problems for public institutions, since their implementation requires decisions to develop new campuses as established campuses approach the suggested ceilings. Yet it is important to recognize that new buildings on new campuses are not likely to be more expensive than new buildings on old campuses and may even be less costly if the new campuses are located outside of major metropolitan areas, where construction labor costs tend to be higher than in smaller communities. And land for new campuses may be less costly than land needed for the expansion of older campuses. It is primarily in large central cities that the acquisition of land for new campuses can be extremely expensive and can pose difficult problems associated with using powers of eminent domain to acquire land in the face of community resistance. On the other hand, as we shall suggest

[5] In Illinois, a special committee appointed by the Illinois Board of Higher Education to investigate issues related to institutional size recently recommended that no college or university should be allowed to plan for a growth of more than 1,000 full-time equivalent students per year (44, p. 2).

in Section 7, some of our large central cities are not now adequately served by public two-year and four-year colleges.

For many private institutions, achieving the suggested minimum sizes is likely to be difficult, if not impossible. Small private institutions are often heavily dependent on tuition as a source of income. Yet tuition that is high enough to cover instructional costs per student tends to limit severely their capacity to attract students in competition with low-cost public institutions, especially in view of the fact that their small enrollment tends to be associated with very high costs per student. Thus they operate in a kind of vicious circle—the only way they can meet rising costs is by raising tuition, and each tuition increase—even though it may be no greater, percentagewise, than increases being adopted by competing public institutions—tends to widen the dollar tuition gap between small private colleges and their public counterparts. Furthermore, as private colleges and universities grow in size, endowment income per student is likely to decline and heavier reliance must be placed on tuition to cover costs of education per student.

In addition, the Commission does not believe that all private institutions should necessarily seek to reach the suggested minimum enrollments. Just as the goal of a stimulating intellectual environment, rather than cost factors alone, has influenced our choice of suggested maxima, so may that goal be more important than cost factors alone in influencing some excellent private institutions toward decisions not to grow much beyond their present enrollments. There are leading private universities, as suggested earlier, with heavy emphasis on rather specialized research programs, that may wish to retain some of the advantages of small size in relation to their particular objectives. There are leading liberal arts colleges with national reputations that may regard their relatively small size as contributing to a unique campus environment.

On the other hand, there are many private universities and liberal arts colleges that very much need to grow to a size at which they can operate with reasonable efficiency. This is also true of many private two-year colleges which, as a group, appear to be encountering especially acute difficulties in meeting rising costs of education, as we have seen.

The Commission's recommendations for expanded federal aid to needy students and to the institutions in which they enroll, as well as for state aid to private higher education, would help some of the private institutions to expand. However, in many cases, given the financial squeeze in which many private institutions find themselves, expanded federal and state aid is likely to turn out to be essential *simply to allow them to maintain their present enrollments.* In some cases, particularly the numerous cases of private colleges located in relatively small communities and lacking the reputation to attract students from any great distance, the most appropriate answer might be merger with another similar small institution. The Commission has suggested this as an appropriate course for some of the small black colleges in its recent report *From Isolation to Mainstream: Problems of the Colleges Founded for Negroes.* State policies of financial support for private higher education could well be structured to provide financial incentives for merger in appropriate cases.

It is also important to keep in mind that relatively small private institutions may be able to reduce their costs of education per student somewhat, and hence mitigate the financial squeeze in which many find themselves, through curriculum reforms carefully designed to retain some of the advantages of small size while at the same time achieving somewhat greater efficiency and more opportunity for intellectual development of the student. In a recent study, Bowen and Douglass have shown that an "eclectic model" of curriculum reform—combining large classes in some parts of the curriculum with small classes in others and incorporating certain other innovative proposals—would result in a significant reduction in the overall cost of education per student in a typical high-quality liberal arts college with an enrollment of about 1,200 (45).

<u>The Commission recommends that state plans for the growth and development of public institutions of higher education should, in general, incorporate minimum FTE enrollment objectives of (1) 5,000 students for doctoral-granting institutions; (2) 5,000 students for comprehensive colleges; (3) 1,000 students for liberal arts colleges; and (4) 2,000 students for two-year (community) colleges.</u>

<u>Similarly, the Commission recommends that state plans should, in general, incorporate maximum FTE enrollment objectives of (1) about 20,000 students for doctoral-granting institutions; (2) about 10,000 students for comprehensive colleges; (3) about 2,500 students for liberal arts colleges; and (4) about 5,000 students for two-year (community) colleges.</u>

The Commission also recommends that, in developing their policies for state aid to private institutions, states should study and adopt policies providing financial incentives for expansion in those cases in which private institutions are clearly much too small for efficient operation, but that state policies should not be designed to force growth on private institutions of demonstrably high quality which are desirous of retaining unique characteristics associated with their comparatively small size. In some states it may be desirable, also, to study and adopt policies providing financial incentives for merger of very small private institutions in appropriate cases. The federal government should also encourage small institutions to grow by giving them priority in the awarding of construction grants and by aiding them through its "developing institutions" program.

CLUSTER COLLEGES In a search for ways of retaining some of the advantages of small size within the overall structure of a larger institution, a growing number of American colleges and universities are experimenting with the development of cluster colleges. The inspiration for the concept of cluster colleges comes chiefly from Oxford and Cambridge universities in England—the concept of a cluster of semi-autonomous colleges within the overall framework of a large university. But the earliest examples of cluster colleges in the United States took the form of federations of closely linked colleges—the Claremont Colleges in Southern California and the Atlanta University Center, both of which began to function in this manner in the 1920s. Other relatively early examples of innovations achieving a good many of the advantages of cluster colleges were the "houses" developed at Harvard and Yale. And the most common examples of cluster colleges take the form of one or two small colleges within a larger university or college which is predominantly an institution of a more conventional type, such as experimental Monteith College within Wayne State University.

Probably the institution most closely patterned after the Oxford and Cambridge models is the University of California at Santa Cruz, which was planned as a cluster-college campus of the university. Its first college was established in 1965 and its fifth college in 1969. Its original plan called for one new college each year until there were 20 colleges in all, but it now appears that financial stringency may result in somewhat slower development.

The colleges at Santa Cruz are not separate legal entities like the colleges at Oxford and Cambridge but are subcolleges within a campus of a larger legal entity, the University of California system. Most of the cluster colleges in the United States, except for those that are parts of a federation, are likewise not separate legal entities but are subcolleges of a single university or college.

By 1969, there were 24 colleges and universities in the United States with subcolleges (Appendix B, Table 9). Under their auspices, 43 subcolleges had been established. In addition, there were 13 colleges that were parts of federated systems (Appendix B, Table 10). All the subcolleges had been established since 1959 and the great majority of them since 1965. There is a good deal of evidence to suggest that the acceleration of the movement from 1965 on was at least partly stimulated by a growing feeling, in an environment of increasing campus unrest, that large campuses had indeed become too unwieldy and impersonal to provide a meaningful educational experience to many of their students (46). But it was undoubtedly also stimulated by the burgeoning enrollments of the 1960s, which resulted in a rapid increase in the number of campuses with many thousands of students.

By no means are all the subcolleges, however, parts of universities or colleges with very large numbers of students. To be sure, those at Wayne State University, the University of California at San Diego and at Santa Cruz, Michigan State University, the University of Michigan, Western Washington State College, CUNY, Rutgers, the University of Nebraska, SUNY College at Old Westbury, and Sonoma State College are all parts of large public campuses or systems. But most of the remaining parent institutions of subcolleges are private, and some are quite small liberal arts colleges. Thus a growing urge to experiment, as well as a movement to avoid the perils of great size, has quite evidently played a role.

Cluster colleges vary greatly in size and in the enrollment ceilings that they have set. Quite clearly there are differing perceptions of what constitutes a desirable degree of smallness, with subcolleges that are parts of large institutions tending to have higher prospective enrollment ceilings than those associated with smaller universities or colleges. They also vary in faculty structure, organization of the curriculum, and many other ways. But they tend to have certain characteristics in common. They seek to bring faculty and students into closer contact through emphasis on small classes, informal discussion groups, guided individual instruction, and other approaches designed to get away from the impersonal atmosphere of the large campus. They tend to attract faculty members who are interested in breaking away from traditional curricula and methods of instruction and students who tend to differ from those of their parent institutions primarily in their interest in new experiences and in their openness to the possibility of change (46, chap. 4). They aim at a broad liberal arts education for their students and an intellectual environment that will teach students to think critically and develop a sense of social responsibility.

Some, like the cluster colleges at Santa Cruz and at Michigan State University, are designed as a group of colleges with differing subject matter emphases within a broad liberal arts curriculum. Thus Cowell College at Santa Cruz stresses relationships and unities of humanistic studies, Stevenson College is oriented toward the social sciences, and Crown College is chiefly concerned with science and its effect on human life (47, p. 222). At Michigan State, James Madison College emphasizes the study of public policy problems; Justin Morill College, international and cross-cultural studies; and Lyman Briggs College, the relation of science to man and society (47, p. 731).

The cluster college retains the advantages of smallness while drawing on the resources of its parent institution to achieve both breadth and economies of scale that are not readily available to the small, independent liberal arts college. At Santa Cruz, faculty members hold dual appointments in a college and a campuswide department which offers courses open to students at all colleges. Thus the departments can have more members representing more specialties within a discipline than would be feasible in a small, independent liberal arts college. And, even where the faculty of a cluster college is largely a separate entity, there are usually arrangements for students to take advanced courses offered on the parent campus. At the University of Michigan and at Michigan State, the academic departments lend faculty to subcolleges for limited periods of years, after which they return to their departments.

In addition to sharing campuswide faculty resources, subcolleges may have their own small libraries, but their students also have access to the main campus library system. Similarly, laboratory and computer facilities may be shared.

Along with these advantages, cluster colleges have their problems. As Gaff has pointed out (46, chap. 10), it is difficult to sustain the initial atmosphere of enthusiasm and optimism which tends to characterize the early years of a cluster college. Disagreement over curriculum issues tends to develop as time goes on, especially if some of the experimental approaches are not working well. Often, the initial enthusiasm revolves around the personality and drive of a charismatic leader. If, after a time, the leader leaves, the effect may be very damaging to morale. On the other hand, there have been some situations in which the failure of such a leader to leave has had equally unfortunate results. Some faculty members may find themselves losing their initial enthusiasm if devotion to innovation in undergraduate teaching prevents them from publishing enough to get promoted. And perhaps one of the most serious problems, at least in a few of the cluster colleges, is the complete absence of a core curriculum or any structure whatever in the curriculum as a result of pressure of students for removal of all requirements.

Another major problem is costs. Small classes and individualized instruction are inherently expensive, and thus at least some of the cluster colleges have a number of large lecture classes, especially at the introductory level, along with smaller classes, to keep their costs per student from rising too much above those on the parent campus or on other campuses of the university.

Yet much of the evidence on the experience of cluster colleges thus far is favorable, and, although it is too early to discern the ultimate impact of the cluster college movement on American higher education, the speed with which it has developed in the last five years or so suggests that the cluster colleges of today may be the pioneers in a development which is destined to become much more widespread. We believe that universities, colleges, and state planning bodies should carefully consider this path to avoiding the inflexibilities associated with excessive bigness when they are planning new campuses or the further development of existing campuses. But we also believe that there is a need for continuing evaluation of the experience of cluster colleges, so that in time there will be a solid body of information on what patterns of organization and curriculum structure seem to be best designed to achieve their aspirations.

The Commission recommends that universities, colleges, and state planning agencies carefully study and adopt plans for the development of cluster colleges. The Commission also recommends that the federal government, the states, and private foundations make funds available for research evaluating the comparative experience of these colleges.

FEDERATIONS AND CONSORTIA

Some of the advantages of cluster colleges can be achieved another way—through federation or consortium arrangements among existing institutions. We have already referred to the Claremont Colleges and the Atlanta University Center as the leading examples of federations. Consortia tend to be looser arrangements under which two or more institutions enter into agreements to sponsor specialized programs, to share computer or library facilities, or to offer students a broader range of courses. Only a few existing consortia actually bear much resemblance to cluster colleges, but many more could come to have some of the features of cluster colleges through more extensive arrangements for sharing courses and facilities.

Consortia developed very rapidly during the 1960s, but there is as yet little agreement on the number of consortia in existence. A 1965–66 study indicated that there were about 1,300 consortia (48), but the findings of that study were questioned by another author who contended that more than half of the 1,300 consortia were bilateral arrangements, largely single-purpose, and lacking formal organization or independent government. He went on to identify consortia that were fully institutionalized—that is, those that were tightly organized, legally incorporated, cooperating entities, with their own officers, independent budgets, representative policy-making bodies, and joint programs distinguishable from the programs of constituent members. He also included only voluntary forms of cooperation, excluding statutory bodies such as state coordinating councils. On that basis, he maintained, there were less than 100 consortia in 1967 (49). A more recent article, published early in 1971, indicated that there were 61 consortia that were substantial enough to have both a full-time administration and two or more academic programs requiring long-term financial commitment, and to also include three or more institutions. These 61 consortia included as members a total of over 550 private and public colleges and universities, large and small. And, as evidence of the rapid recent growth of the movement, 12 of the consortia were organized in 1970 (50).

The Claremont Colleges consist of five undergraduate colleges—Claremont Men's College, Harvey Mudd College, Pitzer College, Pomona College, and Scripps College—and a central coordinating institution, Claremont University Center, which conducts the Claremont graduate education program. Each of the undergraduate colleges cooperates in supporting university facilities at the Center but has its own educational emphasis, administration, faculty, students, and campus. Among the shared facilities are a large central library, auditorium, theater, counseling center, and center for continuing education. Total enrollment in the Claremont Colleges in 1970 was about 4,600, of whom 840 were graduate students at the Claremont University Center.

The Atlanta University Center consists of six institutions originally founded for Negroes—Atlanta University, Clark College, the Interdenominational Theological Center, Morehouse College, Morris Brown College, and Spelman College. The six institutions are independently organized and operated, but students at each school have access to resources and facilities of all member schools, including the library of Atlanta University. Some faculty members are shared by several colleges, and the schools have arrangements for student and faculty exchanges. The University Center sponsors concerts and other cultural events. The Martin Luther King, Jr., Memorial Center was established at Atlanta University following the assassination of Dr. King. It includes the Martin Luther King Library and has plans for the development of other cultural activities, including an institute of nonviolent social change and an ecumenical worship center.

Cooperative arrangements among American colleges and universities, as already suggested, vary widely in degree of formality, purpose, and source of funds. Some are primarily research oriented and are designed to provide or make optimum use of highly sophisticated or costly facilities (51). An example is the University Corporation for Atmospheric Research, composed of several institutions centered in Boulder, Colorado, which operates the National Center for Atmospheric Research. A second type of cooperation, of which there are a number of examples, is an arrangement under which a relatively strong institution helps a developing school, as the University of Michigan aids Tuskegee Institute. These two types of cooperative arrangements may involve institutions that are widely separated geographically.

But a third, and especially significant, development is the formation of consortia or cooperative arrangements among institutions within the same metropolitan area or within a broader region for the purpose of sharing faculty, courses, and educational facilities. An example is the Consortium of Universities of the Washington Metropolitan Area, Inc., with an arrangement under which a graduate student at any member school may select from combined offerings of all member schools the courses best meeting his needs. The five institutions involved are The American University, Catholic University of America, Georgetown University, George Washington University, and Howard University—all located in the District of Columbia, all private, and all with relatively modest Ph.D. programs. Among other existing consortia, a few examples of groups of institutions that have developed quite extensive arrangements for sharing instructional facilities at either the graduate or undergraduate level are the following:

Association for Graduate Education and Research in North Texas
Seven colleges and universities, all private, in the Dallas–Fort Worth area, with a variety of cooperative programs for improvement of graduate and undergraduate teaching and research.

Committee on Institutional Cooperation, Evanston, Illinois
The Big Ten universities and the University of Chicago, with an arrangement under which Ph.D. candidates may enroll for courses at any member institution.

Kansas City Regional Council for Higher Education
Sixteen institutions, public and private, in the Kansas City area, with student exchange programs, a library depository, a pilot computer program, and science seminars.

University Center in Georgia, Athens, Georgia
Nine colleges and universities, public and private, with a program of cooperation in library services, departmental conferences, visiting scholars, faculty research, and teacher training. Six of the member institutions are in Atlanta, two are in Decatur, and the ninth is the University of Georgia in Athens.

Connecticut River Valley Colleges, Massachusetts
Amherst, Mount Holyoke, Smith, and the University of Massachusetts have had a cooperative program for some years, under which they conduct a wide range of four-college programs, including a cooperative doctorate. More recently they have sponsored the establishment of experimental Hampshire College, which admitted its first students in the fall of 1970 (52).

Another successful but less formal cooperative arrangement includes Bryn Mawr, Harvard, and Swarthmore colleges, whose full-time students may take any course at any one of the three colleges, thereby benefitting from a broader selection of courses than could be offered by their own institutions alone. Haverford and Swarthmore students, as well as Bryn Mawr graduate students, may also take courses at the University of Pennsylvania. The agreement between Bryn Mawr and Haverford, which are located only about a mile apart, is especially productive. With undergraduate student bodies of about 800 at Bryn Mawr and 700 at Haverford, there are approximately 1,700 registrations for courses at the companion institution in both directions. In addition, Bryn Mawr students may major in several fields offered at Haverford but not at Bryn Mawr, and vice versa.

Many colleges are involved in cooperative arrangements with other institutions to provide for a junior year abroad, summer study abroad, or other opportunities for foreign study. In fact, this is one of the oldest and best-established forms of cooperation.

In view of the recent acceleration of the formation of consortia and other cooperative arrangements, it is very difficult, as in the case of the cluster college movement, to predict their ultimate impact on the future of American higher education. Yet the movement is full of promise. It represents a way in which small liberal arts colleges can strengthen and broaden their programs by sharing facilities and courses with other institutions. It represents a path which universities with modest graduate programs can follow to provide for their graduate students a much broader selection of advanced courses than could be offered by any one of the cooperating institutions. And, as the Committee on Institutional Cooperation illustrates, it can even be used by large universities with comprehensive doctoral programs to improve access to highly specialized courses.

Nevertheless, despite the promise in the consortium movement, many existing consortia are largely arrangements on paper that have little actual impact. Universities and colleges tend to be reluctant to relinquish their own sovereignty in program development and aim continuously at strength in all fields. Such policies are extremely shortsighted when there are major potentialities for sharing resources and facilities with neighboring institutions or, in some cases, "not so neighboring" institutions.

The Commission recommends that colleges and universities continue to seek ways of sharing facilities, courses, and specialized programs through cooperative arrangements; that existing consortia make continuous efforts toward increasing the effectiveness of their cooperative programs; and that institutions—especially small colleges—that are not now members of consortia carefully consider possibilities for forming consortia with neighboring institutions.

The Commission reiterates its recommendation in *Quality and Equality* that the proposed National Foundation for the Development of Higher Education aid in planning liberal arts centers to be established by groups of colleges for the purpose of increasing quality, scope, and diversity of undergraduate education; of stimulating more economical and effective use of administrative and teaching personnel; and of sharing library and computer facilities.

PRESERVING AND ENCOURAGING DIVERSITY

Despite the pressures for conformity in American higher education, the strong tendency for comprehensive colleges and emerging universities to want to model their programs after those of the leading research universities, and the recent evidence (41) that institutions of higher education are growing more alike, there are also encouraging signs of recent developments that may play an exceedingly important role in preserving and encouraging diversity in the coming decades. The cluster college movement is certainly one of these. So is the consortium movement, particularly because it may help small liberal arts colleges to survive and at the same time broaden opportunities for their students. The consortium movement could also be an effective mechanism for encouraging the sharing of resources.

However, as we have seen, the very survival of many private institutions in the United States is in serious jeopardy. The increased federal aid that we have recommended as well as state aid to private institutions will be crucial factors in determining whether private institutions can maintain something approximating their present share of enrollment. And, by the same token, they will be crucial factors in helping to preserve diversity. Private colleges and universities have long played an important role in developing innovative and experimental programs, and in providing freedom for their faculty members to explore novel ideas. Their disappearance, as we have suggested in an earlier section, would make more difficult the preservation of academic freedom and innovative educational approaches in public institutions.

In framing its proposals both for federal aid and state aid to private institutions, the Commission has placed its major emphasis on grants for students, with accompanying cost-of-education supplements to the institutions they choose to attend. One of our most important reasons for preferring this approach, rather than other forms of institutional aid, is that we believe it will help in the preservation of diversity. It will enhance the freedom of choice of students among institutions and their ability to attend innovative institutions if they wish.

We also believe that the recommendations we have made in *Less Time, More Options: Education Beyond the High School* for encouraging stop-out periods for students and participation of adults in higher education at various times over the course of their lives will contribute to greater diversity in the ages and characteristics of college students and in their life-styles. Some of these ideas will be discussed more fully in Section 8.

7. Needs for New Institutions

INTRODUCTION Whether or not an individual will enter college often depends on the availability of a relatively open-access institution within commuting distance of his home. It is true that the increased student aid that the Commission has recommended in *Quality and Equality: Revised Recommendations, New Levels of Federal Responsibility for Higher Education* would enable relatively more young persons to attend college away from home. But for many young people, especially in inner cities, attending college away from home may not be feasible even if they can qualify for grants up to $1,000 a year or more and for loans. Available grants will usually not cover all the costs of attending college away from home, and young persons from low-income families are often reluctant to assume indebtedness. For many whose families lack any capacity to contribute to the costs of their college education, part-time work while living at home is likely to seem the only feasible way of supplementing any grants or scholarships for which they may qualify. Young people brought up in ghetto areas, moreover, may frequently be wary of problems of adjusting to the environment of a residential college away from home. And finally, the very existence of an open-access public institution not too far from home may be a significant factor in stimulating young people from families with little educational background to enter college. Whatever the combination of reasons may be, we noted in Section 2 that a considerably larger proportion of high school graduates go on to college if there is a public junior college in the community than if there is no college in the community.

We have observed, also, that young people who live in suburban areas are more likely to attend college than those living in inner cities or in nonmetropolitan areas, and that those living in the poverty portions of large metropolitan areas are especially unlikely to attend college.

Another highly important consideration is the accessibility of adult education programs, including evening classes. These are likely to be especially needed in the central cities of large metropolitan areas.

For all these reasons, the Commission believes that, in planning for new institutions of higher education, special emphasis must be placed on an adequate supply of relatively open-access colleges in large urban areas. We do not believe that there is a need for any new doctoral-granting institutions, for reasons that will be discussed in the Commission's forthcoming report *College Graduates and Jobs*. But we do believe that there is a need, especially in large metropolitan areas, for an adequate supply of relatively open-access comprehensive colleges and community colleges, especially when existing universities in the area are national, rather than regional, universities, or are so large that they should not be encouraged to grow.

In our report *The Open-Door Colleges: Policies for Community Colleges*, we recommended the establishment of 230 to 280 new public community colleges by 1980. We also presented state-by-state estimates of needs for new community colleges within the national total. As we pointed out in that report:

The Commission believes that there should be a community college within commuting distance of every potential student, except in sparsely populated areas where residential community colleges should be developed. . . .

Implementation of the Commission's recommendations for new community colleges by 1980 would mean that more than 95 percent of all potential students would live within commuting distance of a community college.

Between 1968 and 1970, there was a net addition of about 80 new institutions to the nation's supply of public two-year colleges, but only 25 of these were in large metropolitan areas. After reviewing needs on a state-by-state and metropolitan area basis, we have revised our nationwide estimate of needs for new community colleges to 175 to 235. In this report, we shall be concerned with the question of how many of these additional community colleges should be located in large metropolitan areas, and we will also consider needs for new comprehensive colleges.

NEEDS FOR NEW URBAN INSTITUTIONS In developing estimates of needs for new urban institutions, the Commission has assembled data on the supply of existing institutions in the central cities and outer rings of large metropolitan areas, along with data on their enrollment, control, and selectivity. The analysis of needs has been carried out for all metropolitan areas with populations of 500,000 or more in 1970, but we have also compiled data on the supply of institutions in smaller metropolitan areas and in nonmetropolitan areas.

Our estimates of needs for new urban institutions in large metropolitan areas are presented on Maps 1 and 2 and in Appendix B, Table 11. They show, in rounded numbers, a need for approximately 60 to 70 new comprehensive colleges and 80 to 125 new community colleges by 1980. The 80 to 125 new community colleges recommended for these areas are not an addition to our estimates of a need for 175 to 235 new community colleges but are included within that estimate.

The data in Appendix B, Table 11 indicate that enrollment rates —in this case, total enrollment in existing institutions in the area as a percentage of population—vary widely by metropolitan area and that certain regional patterns may be discerned in the differences. To aid in the interpretation of differences, the areas have been classified by their enrollment rates (Appendix B, Table 12).

Among the 12 areas with relatively high enrollment rates of 5.2 percent or more, six were in California, with its highly developed community college system, in which 61 percent of all undergraduates in the state were enrolled in 1968. But the California areas in this group all had, in addition to their community colleges, one or more research universities and one or more large state colleges. Phoenix, Arizona, owed its high enrollment rate to the presence of Arizona State University within the Phoenix Metropolitan Area and to Arizona's well-developed community college system, with four community colleges in the Phoenix area in 1968; Boston, largely to its wealth of long-established national and regional private universities; Akron and Columbus, to the presence of large state university campuses; Oklahoma City, to the presence of both the University of Oklahoma and Central State College in its outer ring; and Syracuse, primarily to the presence of Syracuse University, although the area also included a number of other institutions, both public and private.

The second highest group, with enrollment rates of 4.2 to 5.2 percent, also included a combination of eastern areas with long-established private universities and of midwestern or western areas in states with well-developed public systems. Albany-Schenectady-Troy, however, was an eastern area which fell into the group because of a strong combination of public and private

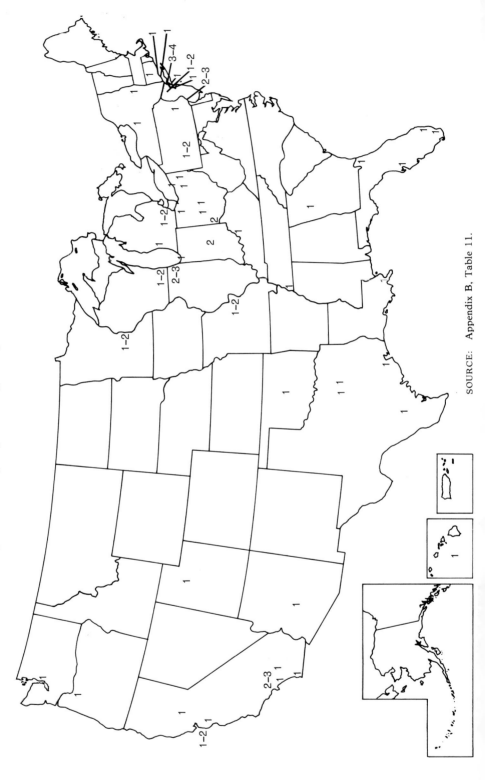

MAP 1 Estimated needs for new comprehensive colleges by 1980 in large metropolitan areas

SOURCE: Appendix B, Table 11.

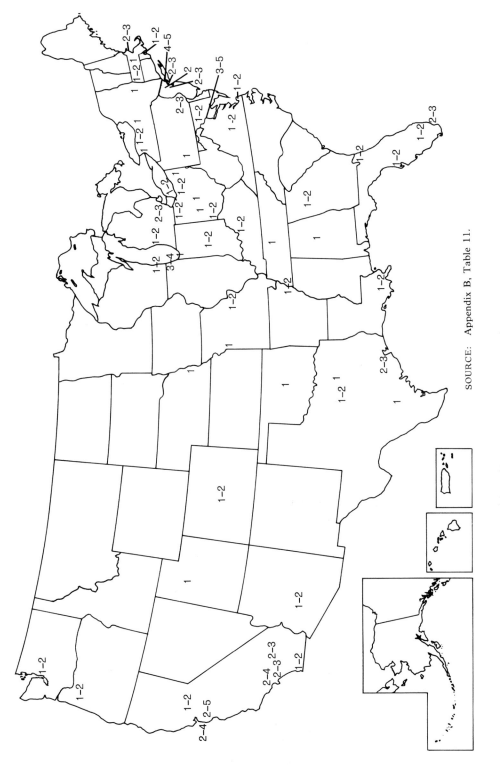

SOURCE: Appendix B, Table 11.

institutions. So did the Washington, D.C., Metropolitan Area, which included the main campus of the University of Maryland in its suburbs, along with the five private universities in the District of Columbia.

The middle or average group, as might have been expected, included areas in nearly all parts of the nation. This was also true of the next group, with enrollment rates of 2.2 to 3.2 percent, but one could discern certain patterns of reasons for deficient enrollment in this group of areas—most had large minority-group populations; 9 of the 23 were in the South; two were in New Jersey, with its long-standing record of meager state financial support of higher education; and a good many (Baltimore, Chicago, Cleveland, Milwaukee, St. Louis, Birmingham, Atlanta, New Orleans, Memphis, and Houston) had the largest populations of any areas in their respective states, but the main campus of the leading state university was in a smaller community. Detroit is also the largest metropolitan area in its state, with the two leading state universities—the University of Michigan and Michigan State—located in smaller communities, but it does have Wayne State University, which had an enrollment of about 33,000 in 1968. One factor that may play a role in explaining Detroit's relatively low enrollment rate is that, of the eight public community colleges in the area in 1968, seven were located in the outer ring and only one in the central city. In Atlanta and Houston, two of the largest metropolitan areas in the South, all the public community colleges were located in the outer ring.

The lowest group, with enrollment rates of less than 2.2 percent, included one New Jersey area, two Indiana areas, one Missouri–Kansas area, and three in the Southeast states.

In developing our estimates of needs for new comprehensive colleges and community colleges in these large metropolitan areas, we considered these factors:

1 Existing enrollment rates
2 The distribution of enrollment by type of institution and control
3 Total population and the annual average rate of population growth from 1960 to 1970
4 The percentage of the population aged 18 to 21 in 1970
5 Whether existing institutions had enrollments that were below, within, or above our suggested optimum size ranges

6 The probability that existing private institutions, though serving significant needs, would not grow as rapidly as similar public institutions in most instances

7 Measures of selectivity developed by the American Council on Education and by Willingham, as well as the institutions' positions on a tuition scale developed by Willingham (53)

We also examined all available state plans and studies that included estimates of needs for new comprehensive or community colleges. In some cases, however, our estimates indicate a need for more new institutions than do these state plans and studies, partly because we used more restrictive optimum-enrollment ranges.

It should be noted that we are presenting estimates of needs for new urban institutions, partly in order to give an indication of how many are needed nationwide, rather than firm recommendations for their location. Actual decisions about the precise number of new comprehensive and community colleges needed in individual large metropolitan areas should be made by state and local planning agencies.

It should also be emphasized that the new comprehensive colleges and community colleges which we suggest will not serve the purpose of making higher education more accessible to residents of these large metropolitan areas if they do not follow relatively open-access policies. As the Commission pointed out in its report *The Open-Door Colleges:*

The Commission believes that access to higher education should be expanded so that there will be an opportunity within the total system of higher education in each state for each high school graduate or otherwise qualified person. This does not mean that every young person should of necessity attend college—many will not want to attend, and there will be others who will not benefit sufficiently from attendance to justify their time and the expense involved. Thus we favor universal access but not universal attendance in our colleges and universities.

Within the system of higher education, the community colleges should follow an open-enrollment policy, whereas access to four-year institutions should generally be more selective.

However, the Commission believes that public comprehensive colleges should have considerably less selective admission policies than public doctoral-granting institutions.

The policies and functions of urban institutions will be discussed more fully in the Commission's forthcoming report *The Campus and the City*.

There is an important caveat that must be expressed in connection with our estimate of needs for new urban institutions of higher education. If the external degree programs and open universities that are now being initiated prove to be highly popular with students, there may be less need for new institutions of the traditional type. However, as suggested in the previous section, it is impossible to predict the extent to which these innovations will alter existing patterns of participation in higher education. We can only emphasize the importance of continuous evaluation of these experiments and of continuous study of their impact on needs for new institutions by state and local planning bodies.

<u>The Commission recommends that state and local planning bodies develop plans for the establishment by 1980 of about 60 to 70 new comprehensive colleges and 80 to 125 new community colleges in large metropolitan areas with populations of 500,000 or more. Our estimates of needs are presented on Maps 1 and 2 and in Appendix B, Table 11. In determining the location of these new institutions within metropolitan areas, particular emphasis should be placed on the provision of adequate open access places for students in inner city areas.</u>

<u>The Commission also recommends that a special effort be made to develop new institutions in those metropolitan areas which, as indicated in Appendix B, Table 12, have comparatively low ratios of enrollment to population. Especially deficient ratios are found in the following metropolitan areas:</u>

Paterson–Clifton–Passaic, N.J.
Gary–Hammond–East Chicago, Ind.
Indianapolis, Ind.
Kansas City (Mo.–Kans.)

Fort Lauderdale–Hollywood, Fla.
Jacksonville, Fla.
Louisville (Ky.–Ind.)

<u>However, in view of the very recent movement to establish external degree programs and open universities, the Commission recommends that state and local planning bodies continuously study the impact of these innovations on patterns of enrollment and modify estimates of needs for new institutions accordingly.</u>

OTHER NEEDS FOR NEW INSTITUTIONS

The new comprehensive colleges recommended for large metropolitan areas will not meet all the needs for additional comprehensive colleges by 1980. We estimate that, in addition to the 60 to 70 new comprehensive colleges needed in metropolitan areas with populations of 500,000 or more, there will be a need for about 20 to 35 additional comprehensive colleges in somewhat smaller areas, generally those with populations of 200,000 to 500,000 in 1968. This brings our estimate of total needs for new comprehensive colleges by 1980 to about 80 to 105. Estimates of total state-by-state needs are presented on Map 3 and in Appendix B, Table 13.

It should be emphasized that the new comprehensive colleges are needed for the purpose of providing adequate geographic distribution so that they will serve the needs of areas not now adequately served and of those with rapidly growing populations. They are not needed just to absorb projected enrollment increases on a nationwide basis. If we were simply to consider the problem on a nationwide basis, we would find that existing public comprehensive colleges could absorb about four-fifths of our projected increase in enrollment for this group of institutions by 1980 if all those with enrollments below 5,000 were to grow to at least that size. But in the absence of new comprehensive colleges, many metropolitan areas would be left without an adequate supply of places for students. And existing institutions that are located in states or communities with slowly growing or declining populations have little or no capacity for growth. Our estimates suggest that there are about 20 states that need no new comprehensive colleges by 1980. They are mostly sparsely populated states west of the Mississippi River or states in the northeastern or southeastern parts of the country that have been experiencing net outmigration.

It may seem surprising that California, which has a highly developed state college system, is expected to need more new comprehensive colleges than any other state. There are two main reasons for this. First, many of the existing state colleges have very large enrollments—for example, San Diego State College had an enrollment of 34,800 and San Jose State College an enrollment of 33,600 in 1970. In metropolitan areas where existing state colleges are much larger than our suggested maxima, new colleges or new campuses of existing colleges should be established to meet future

MAP 3 *Estimated needs for new comprehensive colleges by 1980, by state*

enrollment increases. Secondly, California has had a very rapid rate of growth, and, although net in-migration to the state has been slowing down in recent years, in terms of sheer numbers California can be expected to have very large additions to its population even if the rate of growth in the 1970s is well below that in the 1960s. Similar reasons explain the need for many new community colleges in California in the 1970s.

Some of the states in which many of the existing state colleges are almost entirely devoted to the training of teachers will face a special problem during the 1970s, as the demand for elementary and secondary school teachers declines. Where these institutions are located in large communities, they should be encouraged to move rapidly to broaden their programs to include training for expanding occupations, such as the allied health professions. Where they are located in small communities, and especially where they are located in sparsely populated states, as in Maine or South Dakota, it may be necessary to adopt a state plan for selective development of new programs in these institutions—with a limited number of new programs in any one institution but comprehensive coverage in the state as a whole—or to convert some of them to two-year colleges, or in some cases to phase them out or to merge them with other similar institutions. Recommendations along these lines will be included in the Commission's forthcoming report *College Graduates and Jobs*.[1]

The Commission recommends that state and local planning agencies develop plans for about 80 to 105 new comprehensive colleges, including those already recommended for large metropolitan areas, by 1980. Our estimates of state-by-state needs are presented on Map 3 and in Appendix B, Table 13.

In addition to the 80 to 125 new community colleges recommended for metropolitan areas with populations of 500,000 or more, we believe that an additional 95 to 110 new community colleges should be established by 1980. In determining the location of these new community colleges, particular attention should be paid to the needs of metropolitan areas with populations from

---

[1] Wisconsin has a particularly difficult problem in this respect with its network of two-year teachers colleges located on a county basis, many of them in very small communities. These are to be gradually phased out.

MAP 4 Estimated needs for new community colleges by 1980, by state

200,000 to 500,000. Total state-by-state needs for new community colleges are presented on Map 4 and in Appendix B, Table 13.

The Commission recommends that state and local planning agencies develop plans for the establishment of about 175 to 235 new community colleges in all, including those already recommended for large metropolitan areas, by 1980.

8. Toward More Flexible Patterns of Participation in Higher Education

THE NEED FOR REFORM IN TRADITIONAL TYPES OF ADULT HIGHER EDUCATION

If students are to be encouraged to stop out for a few years after high school or after several years of college and complete their higher education at a later stage, and if participation is to take the form to an increasing extent of part-time study as adults, then it is imperative that existing weaknesses in the organization and structure of adult higher education in the United States be overcome. Despite the rapid expansion and development that has characterized adult higher education in the last few decades, a study recently published by the American Council on Education identified, among others, the following weaknesses (54):

Although the number of faculty assigned to teach full time in continuing education is gradually increasing, the rate of increase is not as great as that of part-time teachers.

There is growing discontent with the reward system among faculty who teach part-time in continuing education. Compensation for teaching evening courses is almost universally lower than that for regular campus courses, and, typically, little credit toward promotion is gained from participation in extension work.

There is growing discontent also with the limitations placed on extension administrators in recruiting and retaining talent from outside the university.

Programs providing graduate training in adult education are increasing, but not as rapidly as the demand for professional workers.

Discontent is growing among administrators of higher adult education over financial policies and practices imposed on them by general university tradition and policy. The most frequently cited causes of discontent are:

There is a double standard in financial policy: except in the Cooperative Extension Service (agricultural extension), adults pay their own way, whereas youth are subsidized. This policy puts the general extension divisions under pressure to skew their programs to money-making activities (large credit and popular noncredit activities) and limits experimental funds.

When state aid is given, it often is loaded with pet projects of pressure groups and too often is ephemeral; it is the first item to be reduced when a new administration enters office.

Fiscal policies and practices of the central administrations, and especially attitudes of budget officers, tend to be geared to the operation of stable programs for full-time students.

Policies for overcoming some of these problems will be recommended by the Commission in its forthcoming report *Education and Life*. In the meantime, it is important to recognize that very recently steps have been taken toward significant innovations in adult higher education in the United States—largely inspired by British models—namely, the development of external degree programs, open universities, and "universities without walls."

EXTERNAL DEGREES AND THE OPEN UNIVERSITY IN BRITAIN

The University of London has long had an external degree program available to persons who have not completed or have not participated in its regular, internal degree programs. The candidate qualifies for his degree by examination, regardless of how he has received his preparation—through reading, correspondence courses, relevant work experience, or in other ways.

More recently, an exciting new concept in adult higher education has been launched in Britain—The Open University. Actually, the program is available to college-age students as well as to older adults, but it is an adult program in the sense that it is designed to make it easier for persons who are holding a full-time job to obtain a university degree.

First suggested by former Prime Minister Harold Wilson in 1963, and reputed to be a brainchild of the late Aneurin Bevan, the concept was carefully studied and developed during the 1960s, and The Open University began its first courses in January 1971.[1] Perhaps its most distinctive feature is that it is a carefully structured program, leading to a bachelor's degree, that is not "campus-

---

[1] This discussion of The Open University is based primarily on articles that have appeared in reference 55 over the course of the last four years.

bound." Centers are being established all over the country where students may register for a single course or for a combination of courses that will lead to a degree. When the student registers, he pays a fee and is assigned to a tutor who will guide his work and give him a reading list.

Television courses, carefully planned and broadcast over the British Broadcasting Company's Channel 2 at prime viewing hours, form an important part of the program, but by no means all of it. Public libraries are expected to play a significant role and will have to expand their holdings of books that will be used in the courses. Correspondence courses are to provide the central core of the program. So that students may gain maximum benefits from the television programs, or participate if they do not have a television set at home, viewing centers are being established where tutors will be present to discuss the programs with the students. These are likely to be either in public libraries or in local colleges or universities. In addition, residential weekend and summer school programs are planned.

Degree programs will normally take four years to complete, although exceptional students may require only three, and some may take five or more. There will also be nondegree courses of various kinds.

There are no entrance qualifications, but the standards for completion of the work for a degree are designed to be comparable with those of other British universities.

The program is partially subsidized, with The Open University receiving its funds directly from the Ministry for Education and Science rather than from the University Grants Committee. However, the student's fee covers a considerable proportion of his educational costs, which are expected to be appreciably below those in the established universities.

By August 1970, there were 40,817 applicants, of whom 25,000 were accepted for the first year of the program.[2]

---

[2] Interestingly, the great majority of applicants were in professional or managerial occupations — 10,000 were teachers, 3,000 were "professional people," 1,600 were administrators or managers, and another 1,600 were scientists or engineers. Only about 1,400 were workers at lower levels of the occupational ladder, while 2,500 were housewives. Thus, at least on the basis of the first group of applicants, the program was not destined to serve primarily as a means of assisting working-class people to rise to higher occupational levels, as its originators may have hoped, but was largely to provide further education for persons who already occupied professional or managerial positions.

SIMILAR DEVELOPMENTS IN THE UNITED STATES

In 1960, the University of Oklahoma established its College of Continuing Education to administer a new Bachelor of Liberal Studies program (56). This is usually regarded as the first external degree system in the United States, but it is probably more accurate to characterize it as resembling an "open university" program. Although not usually present on the campus, the students are engaged in supervised independent study, followed by short residential seminars. By the fall of 1970, there were 1,500 students enrolled, of whom 1,200 were classified as active. Interestingly, three-fourths of the students were from outside the state of Oklahoma (57). Subsequently, somewhat similar programs were developed at Goddard College, Syracuse University, and the University of South Florida.

In 1970, some of the largest institutions in the country began to move toward the development of external degree programs and/or open universities. In September of that year the State University of New York, supported by grants from the Carnegie Corporation and the Ford Foundation, launched a bachelor's program for persons who could not enroll for full-time study on a college campus. Students would complete most degree requirements by taking extension and correspondence courses, as well as courses offered by educational television stations (58). The next step was SUNY's announcement in the spring of 1971 that the following September it would open a nonresident college which would offer instruction through correspondence work, television, counseling, and occasional seminars. Students completing the program would be eligible for the previously announced external bachelor's degree or for an external associate of arts degree (59, p. 3).

Early in 1971, the Massachusetts Board of Higher Education approved a plan for an open university, the University of the Commonwealth, that would offer external degrees through television courses, correspondence study, and equivalency examinations. The new university would have no campus of its own but would use facilities at existing state-supported institutions (60). A few months later, The Pennsylvania State University announced that it would launch an external degree program (61). At about the same time the President's Task Force on the Extended University at the University of California recommended the development of a "part-time" degree system which had some of the elements of an open university program (62).³ The California state college system is also developing plans for encouraging opportunities for part-time study and for some of the changes in degree structures recommended by the Commission in *Less Time, More Options.*

Another important development in early 1971 was the joint announcement of the College Entrance Examination Board and the Educational Testing Service that they were establishing a 21-member national Commission on Nontraditional Study to review opportunities for college-level learning outside of school and make recommendations concerning the recognition of such study, including the awarding of external degrees in the United States (63).

Somewhat different in concept was the program for the development of "universities without walls" which was announced by the U.S. Office of Education in late 1970. Under this program, planning and pilot grants totaling $365,000 were awarded to 17 colleges and universities for experimental programs that would do away with the tradition of a fixed campus and a fixed-age student body. Students would enroll at particular institutions sponsoring the program, but regular class work would be "mixed with large doses" of noncampus learning experiences, including research assistantships, internships with government and social agencies, travel or service abroad, independent study, and other kinds of field work and seminars. Each institution would form an "adjunct faculty" for the program, which would include government officials, businessmen, scientists, artists, and other professionals outside the academic community.⁴

Meanwhile, strong interest was being displayed at many institutions in improving opportunities for students to stop out or to combine work experience with part-time study. In the spring of 1971, it was reported that 125 Yale students, 51 Radcliffe students, and 1 out of every 15 Harvard students were on "leaves of absence"

---

³ The Task Force reported that part-time degree programs were already operating at Harvard, Columbia, Syracuse, Claremont, the University of Chicago, the City College of New York, and Brooklyn College.

⁴ The institutions participating are Antioch College, Bard College, Chicago State College, Friends World College, Goddard College, Howard University, Loretto Heights College, the University of Massachusetts School of Education, the University of Minnesota, New College at Sarasota, Northeastern Illinois State College, Roger Williams College, Shaw University, Skidmore College, the University of South Carolina, Staten Island Community College, and Stephens College (64).

(65). Hobart and William Smith Colleges announced a plan for deferred entrance, as did Beloit College (66) and (67), and a good many other institutions were reported to be developing deferred entrance and/or leave of absence policies.

It is obviously much too early to do more than speculate about the extent to which these innovations will change future patterns of participation in higher education. No doubt many college-age persons and older adults will welcome the opportunity to participate in the open university type of program, while some college-age persons will take advantage of improved opportunities for stopping out and returning later. But there will almost certainly continue to be many college-age young people who will desire the campus experience—the experience of "going to college," of associating for a prolonged period with college friends of their own age. And many will continue to want to attend college away from home to "get out from under" parental influence or domination.

If stopping out and part-time study are to be encouraged, employers will have to cooperate by reversing the trend toward requiring a college degree for increasing proportions of jobs. This will not be altogether easy, because the trend is attributable in part to the increase in the relative importance of jobs requiring professional or technical training and not entirely, as is sometimes loosely suggested, to increasing use of a bachelor's degree as a form of "certification."

To some extent, also, stopping out and returning to higher education at a later point runs counter to the normal life cycle as it relates to marriage and childbearing. It is easier to finance a college education when one is single or married without children than when there are young children to be cared for and supported. Thus, if existing patterns are really to change greatly, new forms of financing higher education for adults, such as the "educational security" program recommended by the Commission in its report *Less Time, More Options,* will be required.[5] Another possibility is the development of a sabbatical year as a new form of fringe benefit. There are already provisions for accumulating the right to a given paid period of weeks or months away from the job on

---

[5] The "educational security" program would take the form of a new benefit under the social security system—the right to two years of benefits for the financing of higher education, to be paid through payroll taxes on employers and employees, with the benefits to be available on application after a period of sustained employment (68, p. 21).

the basis of years of service in some collective bargaining agreements, such as that of the steelworkers, but these are not related to programs for participation in higher education. It would seem quite feasible to develop sabbatical provisions under which an employee could accumulate the right to paid time away from the job *provided* he used the time to participate in higher education.

Whatever the uncertainties surrounding the future development of the innovations discussed in this section, the Commission believes that they should be encouraged by institutions of higher education, with substantial financial support by government agencies and private foundations. The recent steps in this direction are very encouraging.

The Commission reiterates the recommendations that were made in its report *Less Time, More Options* to encourage more flexible patterns of participation in higher education.

The Commission also recommends that state and federal government agencies, as well as private foundations, expand programs of support for the development of external degree systems and open universities along the lines of programs initiated within the last year or so. It will also be important for governmental bodies and foundations to provide funds for evaluation of these innovative programs as they develop.

9. Summary

In its 335 years of development since the founding of Harvard College in 1636, American higher education has evolved into the most complex system of higher education in the world. In fact, as compared with higher education in many other countries, ours is not a "system" at all but a "universe" of about 2,800 institutions with varying sets of relationships to their own boards of trustees, public and private, and to local, state, and federal government agencies. This universe of institutions of higher education has the great advantage of diversity. As compared with the more monolithic systems of a number of other countries, it also has greater flexibility and capacity to innovate. But some institutions have grown too large to be flexible, while others have remained too small to operate economically or to provide education of high quality to their students.

As we look forward to the growth and development of American higher education in the last three decades of the twentieth century, we believe that a major goal should be the preservation and enhancement of both diversity and flexibility in American higher education, along with the greater equality of opportunity which we have stressed as a goal in earlier reports. And a number of our recommendations would also contribute to the more effective utilization of resources.

To the recommendations we have made in earlier reports, recapitulated in Section 5, the present report adds, and in some cases reiterates, recommendations directed to the following goals.

Preserving and enhancing quality
Avoiding either excessively large or uneconomically small campuses through state plans that incorporate optimum enrollment ranges

Encouraging innovation and experimentation in undergraduate education through expansion of the cluster college movement

Encouraging institutions to achieve more effective use of resources and a broader selection of programs for their students by participation in consortia

Reiteration of recommendations designed to aid and preserve private institutions of higher education

Increasing opportunities for students to attend relatively open-access institutions, especially in inner-city areas, by 1980
Locating about 60 to 70 new comprehensive colleges in metropolitan areas with populations of 500,000 or more

Establishing about 80 to 125 new community colleges in metropolitan areas with populations of 500,000 or more

In determining the location of new community colleges and new comprehensive colleges in these large metropolitan areas, placing particular emphasis on the needs for new institutions in inner-city areas

Establishing about 20 to 35 new comprehensive colleges in smaller metropolitan areas, generally those with populations of 200,000 to 500,000

Increasing opportunities for more flexible patterns of participation in higher education
Developing external degree programs and open universities

Implementing other recommendations included in the Commission's report *Less Time, More Options* designed to encourage more flexible patterns of participation

Revising plans for establishing new nondoctoral institutions and community colleges if external degree systems and open universities alter patterns of participation extensively

Appendix A: Carnegie Commission Classification of Institutions of Higher Education, 1970

This classification includes all institutions listed in the U.S. Office of Education's *Advance Report on Opening Fall Enrollment in Higher Education: Institutional Data, 1970*. Whenever a campus of a multicampus institution is listed separately, it is included as a separate institution in our classification. In a few instances, the Office of Education includes all campuses of an institution in a single listing, and in such cases the institution is treated as a single entry in our classification. Our classification includes 2,827 institutions, as compared with the Office of Education total of 2,565 for 1970. The difference is explained by the fact that, for purposes of obtaining the total number of institutions, we have treated each campus as an institution, whereas the Office of Education treats multicampus systems as single institutions *for purposes of obtaining the total number of institutions.*

Another significant problem arises in connection with medical schools, schools of engineering, schools of business administration, and law schools. These institutions appear separately only if they are listed as separate institutions in *Opening Fall Enrollment.* Most of these professional schools are not listed separately, since their enrollment is included in the enrollment of the parent university or university campus. This is true even in a number of instances in which the professional school is not located on the main campus of the university, but on a separate campus, e.g., Johns Hopkins University School of Medicine.

The classification includes five main groups of institutions and a number of subcategories, or 18 categories in all. They are as follows:

1. Doctoral-Granting Institutions

1.1 Heavy emphasis on research   These are the 50 leading institutions in terms of federal financial support of academic science in at least two of the three years, 1965–66, 1966–67, and 1967–68, provided they awarded at least 50 Ph.D.'s (plus M.D.'s if a medical school was on the same campus) in 1967–68. Rockefeller University was included because of the high quality of its research and doctoral training, even though it did not meet these criteria.

1.2 Moderate emphasis on research   These institutions were on the list of 100 leading institutions in terms of federal financial support in at least two out of three of the above three years and awarded at least 50 Ph.D.'s (plus M.D.'s if a medical school was on the same campus) in 1967–68, or they were among the leading 50 institutions in terms of number of Ph.D.'s (plus M.D.'s if a medical school was on the same campus) awarded in that year. In addition, a few institutions that did not quite meet these criteria, but which have graduate programs of high quality and with impressive promise for future development, have been included in 1.2.

1.3 Moderate emphasis on doctoral programs   These institutions awarded 40 or more Ph.D.'s in 1967–68 (plus M.D.'s if a medical school was on the same campus) or received at least $4 million in total federal financial support in 1967–68. No institution is included that granted less than 20 Ph.D.'s (plus M.D.'s if a medical school was on the same campus), regardless of the amount of federal financial support it received.

1.4 Limited emphasis on doctoral programs   These institutions awarded at least 10 Ph.D.'s in 1967–68, with the exception of a few new doctoral-granting institutions which may be expected to increase the number of Ph.D.'s awarded within a few years.

2. Comprehensive Colleges

2.1 Comprehensive colleges I   This group includes institutions that offered a liberal arts program as well as several other programs, such as engineering, business administration, etc. Many of them offered master's degrees, but all lacked a doctoral program or had only an extremely limited doctoral program. All institutions in this group had at least two professional or occupational programs and enrolled at least 2,000 students in 1970. If an institution's enrollment was smaller than this, it could not be considered very comprehensive.

2.2 Comprehensive colleges II This list includes state colleges and some private colleges that offered a liberal arts program and at least one professional or occupational program, such as teacher training or nursing. Many of the institutions in this group are former teachers colleges which have recently broadened their programs to include a liberal arts curriculum. Private institutions with less than 1,500 students and public institutions with less than 1,000 students in 1970 are not included, even though they may offer a selection of programs, because, with small enrollments, they cannot be regarded as comprehensive. Such institutions are classified as liberal arts colleges. The differentiation between private and public institutions is based on the fact that public state colleges are experiencing relatively rapid increases in enrollment and are likely to have at least 1,500 students within a few years even if they did not in 1970. Most of the state colleges with relatively few students were established quite recently.

3. Liberal Arts Colleges

3.1 Liberal arts colleges—Selectivity I These colleges scored 58 or above on Astin's selectivity index (Alexander W. Astin, *Who Goes Where to College?*, Science Research Associates, Chicago, 1965), *or* they were included among the 200 leading baccalaureate-granting institutions in terms of numbers of their graduates receiving Ph.D.'s at 40 leading doctoral-granting institutions, 1920-1966 (National Academy of Sciences, *Doctorate Recipients from United States Universities, 1958-1966*, Appendix B).

The distinction between a liberal arts college and a comprehensive college is not sharp or clear-cut. Some of the institutions in this group have modest occupational programs but a strong liberal arts tradition. A good example is Oberlin, which awarded 91 Mus.B. degrees out of a total of 564 bachelor's degrees in 1967, as well as 31 M.A.T. degrees out of a total of 41 master's degrees. Its enrollment in 1970 was 2,670. Or, consider two Pennsylvania institutions, Lafayette and Swarthmore. The former awarded 113 B.S. degrees in engineering in 1967 out of a total of 349 bachelor's degrees and has been classified in our comprehensive colleges II group. Its enrollment in 1970 was 2,161. Swarthmore has an engineering program leading to a B.S. degree, but it awarded only 11 B.S. degrees out of a total of 250 bachelor's degrees in 1967 and had a 1970 enrollment of 1,164. Swarthmore has a strong liberal arts tradition and did not meet our minimum enrollment criterion for a private college to be classified as a comprehensive college II in 1970, but our decisions in the cases of Oberlin and Lafayette had to be at least partly judgmental.

3.2 Liberal arts colleges—Selectivity II These institutions include all the liberal arts colleges that did not meet our criteria for inclusion in the first group of liberal arts colleges. Again, in many cases, the distinction between some of the larger colleges in this group and in the comprehensive colleges groups is not sharp and clear-cut, but is necessarily partly a matter of judgment.

In addition, as we pointed out in Section 2, many liberal arts colleges are extensively involved in teacher training, but future teachers tend to receive their degrees in arts and sciences fields, rather than in education.

4. All Two-Year Colleges and Institutes

5. Professional Schools and Other Specialized Institutions

5.1 Theological seminaries, bible colleges, and other institutions offering degrees in religion (not including colleges with religious affiliations offering a liberal arts program as well as degrees in religion).

5.2 Medical schools and medical centers As indicated above, this list includes only those that are listed as separate campuses in *Opening Fall Enrollment*. In some instances, the medical center includes other health professional schools, e.g., dentistry, pharmacy, nursing.

5.3 Other separate health professional schools

5.4 Schools of engineering and technology Technical institutes are included only if they award a bachelor's degree and if their program is limited exclusively or almost exclusively to technical fields of study.

5.5 Schools of business and management Business schools are included only if they award a bachelor's or higher degree and if their program is limited exclusively or almost exclusively to a business curriculum.

5.6 Schools of art, music, design, etc.

5.7 Schools of law

5.8 Teachers colleges Teachers colleges are included only if they do not have a liberal arts program.

5.9 Other specialized institutions Includes graduate centers, maritime academies, military institutes (lacking a liberal arts program), and miscellaneous.

Note: Extension divisions of universities and campuses offering only extension programs are not included. The number of institutions in each of our categories, public and private, may be found in Appendix B, Table 4, while a detailed breakdown of the number of specialized institutions is presented in text Table 1. As noted in the "Foreword," we plan to publish the lists of institutions included in each of the above categories within the next few months.

Appendix B: Tables

TABLE 1 Undergraduate degree-credit enrollment, in numbers and as percent of population aged 18 to 21, United States, actual, 1870 to 1970, and projected, 1980 to 2000

	Undergraduate degree-credit enrollment*				Percent of increase attributable to change in enrollment rate
Year	Number (in thousands)	Percentage change	Percent of population aged 18 to 21	Percent of increase attributable to change in population aged 18 to 21	
1870	52		1.7		
1880	116	122	2.7	30	70
1890	154	33	3.0	65	35
1900	232	50	3.9	29	71
1910	346	49	5.0	34	66
1920	582	68	7.9	10	90
1930	1,053	81	11.9	25	75
1940	1,388	32	14.5	26	74
1950	2,422†	74	26.9	0	100
1960	3,227	33	33.8	19	81
1970	6,840	112	47.6	45	55
1980	10,080	48	59.2	39	61
1990	9,660	—4	67.4	—100	0
2000	12,700	31	72.6	70	30

*Graduate students are included in 1870 and 1880, but their numbers were very small in those years. Most students enrolled in first professional degree programs are included in all years. Before 1960, data are based on the biennial surveys conducted by the U.S. Office of Education, relate to the academic year ending in the designated year, and exclude extension enrollment. From 1960 on, the data relate to opening fall enrollment and include extension enrollment.

†Includes 898,000 veterans of World War II.

SOURCES: U.S. Bureau of the Census, *Historical Statistics of the United States: Colonial Times to 1957,* ser. H 316–326, Washington, D.C., 1960; U.S. Bureau of the Census, *Historical Statistics of the United States: Continuation to 1962 and Revisions,* ser. H 223–350, Washington, D.C., 1965; U.S. Office of Education, unpublished tables; and projections developed by the Carnegie Commission staff, under the direction of Dr. Gus W. Haggstrom of the University of California.

TABLE 2 Graduate degree-credit enrollment, in numbers and as percent of population aged 22 to 24, and total degree-credit enrollment, in numbers and as percent of population aged 18 to 24, United States, actual, 1870 to 1960, and projected, 1970 to 2000 (Projection C for graduate enrollment)

Year	Graduate degree-credit enrollment			Total degree-credit enrollment		
	Numbers (in thousands)	Percentage change	Percent of population aged 22 to 24	Numbers (in thousands)	Percentage change	Percent of population aged 18 to 24
1870	*			52		1.1
1880	*			116	122	1.6
1890	2		0.1	157	35	1.8
1900	6	145	0.1	238	52	2.3
1910	9	57	0.2	355	50	2.9
1920	16	71	0.3	598	68	4.7
1930	47	203	0.7	1,101	84	7.2
1940	106	124	1.5	1,494	36	9.1
1950	237	124	3.3	2,659	78	16.5
1960	356†	50	5.2	3,583	35	22.2
1970	930	161	9.2	7,760	117	31.7
1980	1,570	69	12.6	11,650	50	39.5
1990	1,720	10	15.7	11,380	−2	45.0
2000	1,980	15	16.2	14,680	29	49.4

*Data not available.

†Until 1960, data relate to the academic year ending in the year indicated and were compiled on various bases, e.g., average enrollment during the year; from 1960 on the data relate to opening fall enrollment in the academic year beginning with the year indicated.

SOURCES: U.S. Bureau of the Census, *Historical Statistics of the United States: Colonial Times to 1957*, ser. H 316–326, Washington, D.C., 1960; U.S. Bureau of the Census, *Historical Statistics of the United States: Continuation to 1962 and Revisions,* ser. H 223–350, Washington, D.C., 1965; U.S. Office of Education, unpublished tables; and projections developed by the Carnegie Commission staff, under the direction of Dr. Gus W. Haggstrom of the University of California.

TABLE 3 Birthrate and number of live births, United States, actual, 1910 to 1969, and projected, 1970-71 to 1990-91

Year	Birth-rate*	Live births (in thousands)	Year	Birth-rate*	Live births (in thousands)	Year	Birth-rate	Live births (in thousands)
							\multicolumn{2}{l}{Projections}	
							Series D‡	
1910	30.1	2,777	1949	24.5	3,649			
1920	27.7	2,950	1950	24.1	3,632	1970-71	17.1	3,534
1930	21.3	2,618	1951	24.9	3,823			
1931	20.2	2,506	1952	25.1	3,913	1975-76	18.3	3,959
1932	19.5	2,440	1953	25.0	3,965			
1933	18.4	2,307	1954	25.3	4,078	1980-81	19.3	4,407
1934	19.0	2,396	1955	25.0	4,104			
1935	18.7	2,377	1956	25.2	4,218	1985-86	19.5	4,717
1936	18.4	2,355	1957	25.3	4,308			
1937	18.7	2,413	1958	24.5	4,255	1990-91	18.6	4,765
1938	19.2	2,496	1959	24.0	4,245			
							Series E‡	
1939	18.8	2,466	1960	23.7	4,258			
1940	19.4	2,559	1961	23.3	4,268	1970-71	16.5	3,402
1941	20.3	2,703	1962	22.4	4,167			
1942	22.2	2,989	1963	21.7	4,098	1975-76	17.5	3,766
1943	22.7	3,104	1964	21.0	4,027			
1944	21.2	2,939	1965	19.4	3,760	1980-81	18.0	4,080
1945	20.4	2,858	1966	18.4	3,606			
1946	24.1	3,411	1967	17.8	3,521	1985-86	17.6	4,178
1947	26.6	3,817	1968	17.5	3,502			
1948	24.9	3,637	1969†	17.7	3,571	1990-91	16.3	4,057

*Live births per 1,000 population.
†Preliminary.
‡See footnote 4, p. 44.

SOURCES: U.S. Bureau of the Census, *Historical Statistics of the United States: Colonial Times to 1957*, ser. B 6 and B 19, Washington, D.C., 1960; U.S. Bureau of the Census, *Historical Statistics of the United States: Continuation to 1962 and Revisions*, ser. B 6 and B 19, Washington, D.C., 1965; *Statistical Abstract of the United States, 1970*, p. 47; and U.S. Bureau of the Census, *Population Estimates and Projections*, ser. P-25, no. 448, Washington, D.C., August 6, 1970, p. 9.

TABLE 4
Enrollment in institutions of higher education and number of institutions, by type of institution and control, United States, 1970

Type of institution	Enrollment (in thousands)		
	Public	Private	Total
Total*	6,364.4	2,132.0	8,496.2
Doctoral-granting institutions	1,900.8	636.9	2,537.7
Heavy emphasis on research	774.3	236.5	1,010.8
Moderate emphasis on research	579.2	155.1	734.5
Moderate emphasis on doctoral programs	278.8	135.6	414.4
Limited emphasis on doctoral programs	268.5	109.7	378.2
Comprehensive colleges	2,109.9	533.6	2,643.4
Comprehensive colleges I	1,679.6	411.5	2,091.1
Comprehensive colleges II	430.3	122.1	552.3
Liberal arts colleges	34.8	643.9	678.6
Selectivity I	0.0	156.5	156.5
Selectivity II	34.8	487.4	522.1
Two-year institutions	2,214.0	133.8	2,347.8
Specialized institutions	104.9	183.6	288.5

*Excludes extension enrollment separately reported by institutions; items may not add to totals because of rounding.

SOURCE: Adapted from U.S. Office of Education data by the Carnegie Commission staff.

		Number of institutions				
Percent public	Percent of total	Public	Private	Total	Percent public	Percent of total
74.9	100.0	1,313	1,514	2,827	46.4	100.0
74.9	29.9	101	63	164	61.6	5.8
76.6	11.9	26	20	46	56.5	1.6
78.9	8.6	30	18	48	62.5	1.7
67.3	4.9	23	12	35	65.7	1.2
71.0	4.5	22	13	35	62.9	1.2
79.8	31.1	316	147	463	68.3	16.4
80.3	24.6	210	91	301	69.8	10.6
77.9	6.5	106	56	162	65.4	5.7
5.1	8.0	27	676	703	3.8	24.9
0.0	1.8	0	121	121	0.0	4.3
6.7	6.1	27	555	582	4.6	20.6
94.3	27.6	805	256	1,061	75.9	37.5
36.4	3.4	64	372	436	14.7	15.4

TABLE 5 Undergraduate and postbaccalaureate enrollment in institutions of higher education, by type of institution and control, United States, 1968 (numbers in thousands)

Type of institution	Public				Private
	Undergraduate	Postbaccalaureate	Total	Percent postbaccalaureate	Undergraduate
Total enrollment*	4,772.2	647.2	5,419.4	11.9	1,711.0
Doctoral-granting institutions	1,374.5	376.2	1,750.7	21.5	392.8
Heavy emphasis on research	530.8	177.1	707.9	25.0	126.9
Moderate emphasis on research	418.1	112.8	530.9	21.2	98.0
Moderate emphasis on doctoral programs	225.4	47.1	272.5	17.3	89.9
Limited emphasis on doctoral programs	200.2	39.2	239.4	16.4	78.0
Comprehensive colleges	1,571.0	235.7	1,806.7	13.0	441.4
Comprehensive colleges I	1,263.4	180.4	1,443.8	12.5	331.1
Comprehensive colleges II	307.6	55.3	362.9	15.2	110.3
Liberal arts colleges	21.4	0.6	22.0	2.8	607.4
Selectivity I	0.0	0.0	0.0	0.0	143.5
Selectivity II	21.4	0.6	22.0	2.8	463.9
Two-year institutions	1,745.1	3.2‡	1,748.3	0.2‡	156.5
Specialized institutions	60.2	31.5	91.7	34.3	112.8

*Excludes extension enrollment; items may not add to totals because of rounding.
†Less than 0.05.
‡The presence of small numbers of graduate students in two-year institutions is explained by the fact that there are a few institutions which have predominantly two-year programs, but which also have upper-division and graduate programs in selected fields.
SOURCE: Adapted from U.S. Office of Education data by Carnegie Commission staff.

				Total			
Post-bacca-laureate	Total	Percent post-bacca-laureate	Under-graduate	Post-bacca-laureate	Total	Percent post-baccalaureate
389.9	2,100.9	18.6	6,483.2	1,037.1	7,520.3	13.8
232.9	625.8	37.2	1,767.3	609.2	2,376.5	25.6
116.0	242.9	47.8	657.7	293.1	950.9	30.8
57.9	155.9	37.2	516.1	170.7	686.8	24.9
34.4	124.2	27.6	315.3	81.4	396.7	20.5
24.7	102.7	24.1	278.2	63.9	342.1	18.7
79.7	521.1	15.3	2,012.4	315.3	2,327.8	13.5
69.9	401.0	17.4	1,594.5	250.3	1,844.8	13.6
9.7	120.1	8.1	417.9	65.0	482.9	13.5
24.2	631.6	3.8	628.8	24.8	653.7	3.8
5.5	148.9	3.7	143.5	5.5	148.9	3.7
18.8	482.7	3.9	485.4	19.4	504.7	3.8
0.0†	156.6	0.0†	1,901.6	3.2‡	1,904.8	0.2
53.0	165.8	32.0	172.9	84.5	257.4	32.8

TABLE 6 Undergraduate resident degree-credit enrollment in higher education, percent of population aged 18 to 21, net in-migration or out-migration of students, and enrollment of state residents as percent of population aged 18 to 21, by state, United States, 1968

State	Undergraduate enrollment Number	Percent of population aged 18 to 21	Net in-migration (+) or out-migration (−) of students as percent of enrollment	Undergraduate state residents as percent of population aged 18 to 21
New England states	358,998	48		32
Connecticut	76,731	40	−27	32
Maine	20,108	29	+ 8	18
Massachusetts	196,279	56	+13	38
New Hampshire	21,587	46	+24	22
Rhode Island	28,339	44	+13	28
Vermont	15,954	58	+41	23
North Atlantic states	1,087,879	40		34
Delaware	11,290	30	− 4	18
District of Columbia	36,926	62	+51	17
Maryland	88,117	32	−13	27
New Jersey	122,863	27	−69	24
New York	544,609	47	−10	42
Pennsylvania	284,074	39	− 4	32
North Midwest states	1,103,363	42		35
Illinois	294,376	42	−13	38
Indiana	141,774	42	+14	31
Michigan	249,513	43	+ 5	38
Ohio	280,634	39	+ 3	32
Wisconsin	137,066	50	+11	40
Central states	510,651	47		37
Iowa	85,588	48	+ 5	35
Kansas	80,974	49	+ 8	40
Minnesota	111,218	46	+ 3	39
Missouri	132,590	44	+ 9	34
Nebraska	52,941	51	+13	39
North Dakota	22,411	47	− 2	40
South Dakota	24,929	52	+ 9	41
Southeast states	1,010,392	32		25
Alabama	83,851	33	+ 4	28
Arkansas	45,415	34	+ 2	29
Florida	158,612	37	− 1	32

State	Undergraduate enrollment		Net in-migration (+) or out-migration (−) of students as percent of enrollment	Undergraduate state residents as percent of population aged 18 to 21
	Number	Percent of population aged 18 to 21		
Georgia	90,894	26	+ 4	21
Kentucky	80,155	35	+10	27
Louisiana	97,934	36	+ 4	33
Mississippi	58,408	34	+ 3	30
North Carolina	110,724	29	+18	21
South Carolina	42,632	20	+ 1	15
Tennessee	104,054	37	+17	27
Virginia	88,209	24	−11	18
West Virginia	49,504	41	+20	28
Southwest states	495,457	41		37
Arizona	65,048	52	+ 8	43
New Mexico	32,058	39	+ 1	32
Oklahoma	85,296	50	+ 7	42
Texas	313,055	38	+ 3	35
Mountain states	194,374	55		40
Colorado	78,785	52	+18	37
Idaho	24,124	49	− 4	39
Montana	22,613	44	− 3	38
Utah	57,953	74	+29	51
Wyoming	10,899	48	− 8	39
Pacific states	920,583	48		44
Alaska	2,738	9	−54	8
California	708,614	50	+ 1	47
Hawaii	19,036	28	−16	23
Nevada	8,107	23	−17	19
Oregon	70,509	51	+ 6	43
Washington	111,579	48	+ 4	42

SOURCES: U.S. Office of Education, *Opening Fall Enrollment in Higher Education: Part A—Summary Data, 1968,* Washington, D.C., 1969; estimates of population aged 18 to 21 developed by Carnegie Commission staff; and U.S. Office of Education, unpublished data on residence and migration of students, fall 1968.

TABLE 7 Enrollment in higher education by type of institution, United States, actual, 1970, and projected, 1980-2000, and percent of enrollment in public institutions, 1980, projection C (numbers in thousands)

Type of institution	1970	1980	1990
Total*	8,494	12,977	12,618
Doctoral-granting institutions	2,537	3,471	3,361
Heavy emphasis on research	1,011	1,288	1,244
Moderate emphasis on research	734	1,036	1,004
Moderate emphasis on doctoral programs	414	568	550
Limited emphasis on doctoral programs	378	579	563
Comprehensive colleges	2,643	4,171	4,060
Comprehensive colleges I	2,091	3,283	3,195
Comprehensive colleges II	552	888	865
Liberal arts colleges	678	898	868
Selectivity I	156	194	187
Selectivity II	522	704	681
Two-year institutions	2,348	3,994	3,898
Specialized institutions	288	443	431

*Excludes extension enrollment reported separately by institutions; the data relate to the 50 states and the District of Columbia, excluding outlying territories.

SOURCE: Adapted from U.S. Office of Education data by the Carnegie Commission staff.

2000	Percentage change, 1970–1980	Percentage of total, 1980	Percent public, 1980
16,505	53	100	79
4,205	37	27	79
1,506	27	10	82
1,273	41	8	85
688	37	4	69
738	53	5	74
5,375	58	32	84
4,222	57	25	84
1,153	61	7	83
1,070	32	7	5
223	24	2	0
847	35	5	7
5,291	70	31	96
564	54	3	40

TABLE 8
Undergraduate and postbaccalaureate enrollment, actual, 1970, and three alternative projections, 1975 to 2000 (numbers in thousands)

	Projection A				Projection B	
Year	Undergraduate*	Postbaccalaureate†	Total‡	Percentage change	Undergraduate*	Postbaccalaureate†
1970	7,285	1,213	8,498		7,300	1,198
1975	9,656	1,862	11,518	36	9,656	1,716
1980	11,082	2,441	13,523	17	11,082	2,162
1985	10,544	2,773	13,317	−2	10,544	2,411
1990	10,587	2,749	13,336	§	10,587	2,371
1995	12,288	2,851	15,138	14	12,288	2,462
2000	14,123	3,239	17,361	15	14,123	2,798

*Excludes most candidates for first professional degrees. Actual data for undergraduate and postbaccalaureate enrollment for 1970 are not yet available, although total enrollment is available. Data for postbaccalaureate enrollment presented here are based on our three alternative projections; undergraduate enrollment is the difference between total enrollment and estimated postbaccalaureate enrollment.

†Includes candidates for master's and doctor's, and most candidates for first professional degrees.

‡Items may not add to totals because of rounding. Includes extension enrollment separately reported. The data relate to the 50 states and the District of Columbia.

§Less than 0.5 percent.

SOURCE: Projections developed by the Carnegie Commission staff under the direction of Gus W. Haggstrom of the University of California.

		Projection C			
Total‡	Percentage change	Undergraduate*	Postbaccalaureate†	Total‡	Percentage change
8,498		7,313	1,185	8,498	
11,372	34	9,656	1,587	11,243	32
13,244	16	11,082	1,933	13,015	16
12,954	—2	10,544	2,116	12,659	—3
12,958	§	10,587	2,068	12,654	§
14,750	14	12,288	2,146	14,434	14
16,921	15	14,123	2,436	16,559	15

TABLE 9 Colleges and universities with subcolleges, 1969

Institution	Date of establishment	Approximate enrollment ceiling
Wayne State University, Michigan		
Monteith College	1959	1,200
Wesleyan University, Connecticut		
College of Social Studies	1960	100
College of Letters	1960	100
University of the Pacific, California		
Raymond College	1962	250
Elbert Covell College	1963	250
Callison College	1967	250
University of California, Santa Cruz		
Cowell College	1965	600
Stevenson College	1966	600
Crown College	1967	600
Merrill College	1968	600
College Number Five	1969	600
Goddard College, Vermont		
Greatwood		200
Northwood	1965	200
Hofstra University, New York		
New College	1965	150
Michigan State University		
Justin Morrill College	1965	1,200
Lyman Briggs College	1967	1,200
James Madison College	1967	1,200
Nasson College, Maine		
New Division	1965	
Oakland University, Michigan		
Charter College	1965	350
New College	1967	350
Allport College	1969	350
University of California, San Diego		
Revelle College	1958	2,500
Muir College	1967	1,500
Fordham University, New York		
Bensalem College	1967	100

Institution	Date of establishment	Approximate enrollment ceiling
University of Michigan		
The Residential College	1967	1,200
Western Washington State College		
Fairhaven College	1967	600
City University of New York, Kingsborough Community College		
Brighton	1968	750
Darwin	1968	750
Rutgers—The State University of New Jersey		
Livingston College	1968	3,500
Colby College, Maine		
Program in Human Development	1969	400
Program in Intensive Studies in Western Civilization	1969	400
Program in Bilingual and Bi-cultural Studies	1969	400
Grand Valley State College, Michigan		
Thomas Jefferson College	1969	600
University of Nebraska		
Centennial Education Program	1969	600
State University of New York, College at Old Westbury		
Urban Studies College	1969	500
Disciplines College	1969	500
General Program	1969	500
Redlands University, California		
Johnston College	1969	600
St. Edwards University, Texas		
Holy Cross College	1967	
Maryhill College	1967	
St. Olaf College, Minnesota		
The Paracollege	1969	500
Sonoma State College, California		
Hutchins School of Liberal Studies	1969	750
University of Vermont		
Experimental Program	1969	400

SOURCE: Jerry G. Gaff and Associates, *The Cluster College,* Jossey-Bass, Inc., San Francisco, 1970, pp. 16–17.

TABLE 10
Federated colleges and universities, 1969

Institution	Date of establishment	Enrollment, 1968
Claremont Colleges, California		4,471
Pomona College	1887	1,313
Claremont Graduate School and University Center	1925	830
Scripps College	1926	526
Claremont Men's College	1946	795
Harvey Mudd College	1955	345
Pitzer College	1963	662
Atlanta University Center of Higher Education, Georgia		
Atlanta University	1929*	1,056
Morehouse College	1929*	1,035
Spelman College	1929*	945
Morris Brown College	1932*	1,372
Clark College	1941*	1,003
Interdenominational Theological Center	1958	113
Hamilton College, New York	1793	853
Kirkland College	1968	172

*Date when cooperation began.
SOURCE: Jerry G. Gaff and Associates, *The Cluster College*, Jossey-Bass, Inc. San Francisco, 1970, p. 29; enrollment data from U.S. Office of Education, *Opening Fall Enrollment in Higher Education: Part B—Institutional Data, 1968*, Washington D.C., 1969.

TABLE 11
Estimated needs for new comprehensive colleges and community colleges or new campuses of existing institutions in large metropolitan areas* by 1980

	Population, 1968 (thousands)	Total enrollment † as percent of population, 1968		
		Public	Private	Total
New England states				
Connecticut				
Bridgeport	780	0.77	1.88	2.6
Hartford	800	1.83	1.57	3.4
New Haven	730	1.98	2.31	4.3
Massachusetts				
Boston	3,290	.93	4.39	5.3
Springfield (Mass. and Conn.)	550	1.43	2.65	4.0

Estimated needs for new colleges, 1980	
Comprehensive colleges	*Community colleges*
1	
1	
1	
	2-3
	1

(Table continued on pages 144-151)

TABLE 11
(continued)

	Population, 1968 (thousands)	Total enrollment † as percent of population, 1968		
		Public	Private	Total
Worcester	610	1.28	2.05	3.32
Rhode Island				
Providence	750	1.06	2.47	3.52
North Atlantic states				
District of Columbia				
Washington (D.C., Md., and Va.)	2,750	2.43	2.39	4.82
Maryland				
Baltimore	1,980	1.70	1.38	3.07
New Jersey				
Jersey City	620	1.20	1.20	2.40
Newark	1,880	1.76	1.16	2.92
Paterson-Clifton-Passaic	1,350	0.61	1.22	1.83
New York				
Albany-Schenectady-Troy	710	2.18	2.50	4.68
Buffalo	1,320	3.07	0.81	3.89
New York	11,550	1.83	1.73	3.56
Rochester	850	1.89	2.71	4.60
Syracuse	630	2.25	3.91	6.16
Pennsylvania				
Allentown-Bethlehem-Easton (Pa.-N.J.)	530	0.51	2.23	2.75
Philadelphia (Pa.-N.J.)	4,830	2.09	1.65	3.74
Pittsburgh	2,390	1.72	1.05	2.77
North Midwest states				
Illinois				
Chicago	6,820	1.44	1.54	2.99
Indiana				
Gary–Hammond–East Chicago	610	1.21	0.92	2.13
Indianapolis	1,060	1.20	0.79	1.99
Michigan				
Detroit	4,130	2.44	0.63	3.07
Grand Rapids	510	1.47	1.24	2.71

| Estimated needs for new colleges, 1980 ||
Comprehensive colleges	Community colleges
	1–2
	1–2
	3–5‡
	1–2
1	2–3
1	2–3
1–2	2
1	1
	1
3–4	4–5
	1–2
1	1
1	
2–3	2–3
1–2	1
2–3	3–4
1	
2	1–2
1–2	2–3
1	1–2

**TABLE 11
(continued)**

	Population, 1968 (thousands)	Total enrollment † as percent of population, 1968		
		Public	Private	Total
Ohio				
Akron	680	5.12	0.17	5.29
Cincinnati (Ohio-Ky.-Ind.)	1,380	2.22	0.67	2.89
Cleveland	2,070	1.35	1.16	2.51
Columbus	870	4.76	1.29	6.05
Dayton	840	1.64	2.24	3.88
Toledo (Ohio-Mich.)	680	4.01	0.20	4.21
Youngstown-Warren	530	2.80		2.80
Wisconsin				
Milwaukee	1,340	1.94	0.61	2.55
Central states				
Minnesota				
Minneapolis–St. Paul	1,680	4.00	0.87	4.87
Missouri				
Kansas City (Mo.-Kansas)	1,230	1.38	0.65	2.04
St. Louis (Mo.-Ill.)	2,330	1.71	1.20	2.91
Nebraska				
Omaha (Nebr.-Iowa)	530	2.25	0.99	3.24
Southeast states				
Alabama				
Birmingham	740	1.51	0.83	2.34
Florida				
Fort Lauderdale–Hollywood	530	0.92	0.19	1.11
Jacksonville	510	1.06	0.77	1.84
Miami	1,150	2.10	1.46	3.56
Tampa–St. Petersburg	920	2.68	0.44	3.12
Georgia				
Atlanta	1,330	1.86	0.97	2.83
Kentucky				
Louisville (Ky-Ind.)	800	1.28	0.51	1.79
Louisiana				
New Orleans	1,060	1.16	1.50	2.67

Estimated needs for new colleges, 1980	
Comprehensive colleges	Community colleges
	1–2
2	1–2
1	1–2
1	1
1	1
1	1–2
1	1
1–2	1–2
1–2	
	1
1–2	1–2
	1
	1
1	1–2
1§	1–2
1§	2–3
1	1–2
1	1–2
1	1–2
1	1–2

TABLE 11 (continued)

	Population, 1968 (thousands)	Total enrollment† as percent of population, 1968		
		Public	Private	Total
North Carolina				
Greensboro–Winston-Salem–High Point	590	2.43	1.44	3.87
Tennessee				
Memphis (Tenn.-Ark.)	770	2.93	1.28	4.21
Nashville	540	0.84	1.57	3.41
Virginia				
Norfolk-Portsmouth	650	2.31	0.06	2.38
Richmond	510	2.81	1.34	4.15
Southwest states				
Arizona				
Phoenix	870	6.04	0.17	6.21
Oklahoma				
Oklahoma City	610	5.32	0.99	6.30
Texas				
Dallas	1,460	1.98	1.08	3.06
Fort Worth	680	2.91	1.32	4.22
Houston	1,870	2.45	0.65	3.10
San Antonio	850	1.45	1.10	2.54
Mountain states				
Colorado				
Denver	1,130	3.07	1.09	4.16
Utah				
Salt Lake City	530	3.76	0.46	4.22
Pacific states				
California				
Anaheim–Santa Ana–Garden Grove	1,260	4.86	0.48	5.34
Los Angeles–Long Beach	6,860	4.66	0.69	5.34
Sacramento	760	7.18		7.18
San Bernardino	1,090	3.54	0.78	4.32
San Diego	1,220	6.03	0.40	6.43
San Francisco–Oakland	3,000	5.14	0.44	5.58
San Jose	990	6.89	1.77	8.66

Estimated needs for new colleges, 1980	
Comprehensive colleges	Community colleges
1	1–2
	1
	1–2
	1–2
1	1–2
1	1
1	1
	1–2
1	2–3
1	1
	1–2
1	1
1	2–3
2–3	2–4
1	1–2
1	2–3
1	1–2
1–2	2–4
1	2–3

**TABLE 11
(continued)**

	Population, 1968 (thousands)	Total enrollment † as percent of population, 1968		
		Public	Private	Total
Hawaii				
Honolulu	630	3.46	0.32	3.78
Oregon				
Portland (Ore.-Wash.)	960	2.80	0.93	3.73
Washington				
Seattle-Everett	1,340	4.34	0.46	4.80
Total needs for new colleges in large metropolitan areas				

*Includes all metropolitan areas with estimated population of 500,000 or more in 1968
† Total enrollment in all institutions of higher education in the area.
‡ Additional colleges are needed chiefly in the suburbs.
§ A new university that could serve as a comprehensive college has been established.
SOURCE: Estimates developed by Carnegie Commission staff.

Estimated needs for new colleges, 1980	
Comprehensive colleges	Community colleges
1	
1	1–2
1	1–2
57–68	80–126

TABLE 12 *Enrollment in institutions of higher education in the area as a percentage of population, large metropolitan areas,* United States, 1968*

Less than 2.2	2.2 to 3.2	3.2 to 4.2	4.2 to 5.2	5.2 and over
Paterson-Clifton-Passaic	Bridgeport	Hartford	New Haven	Boston
Gary-Hammond-East Chicago	Baltimore	Springfield	Washington, D.C.	Syracuse
Indianapolis	Jersey City	Worcester	Albany-Schenectady-Troy	Akron
Kansas City	Newark	Providence	Rochester	Columbia
Fort Lauderdale-Hollywood	Allentown-Bethlehem-Easton	Buffalo	Toledo	Phoenix
Jacksonville	Pittsburgh	New York	Minneapolis-St. Paul	Oklahoma City
Louisville	Chicago	Philadelphia	Memphis	Anaheim-Santa Ana-Garden Grove
	Detroit	Dayton	Fort Worth	Los Angeles-Long Beach
	Grand Rapids	Omaha	Salt Lake City	Sacramento
	Cincinnati	Miami	San Bernardino	San Diego
	Cleveland	Greensboro-Winston-Salem-High Point	Seattle-Everett	San Francisco-Oakland
	Youngstown-Warren	Richmond		San Jose
	Milwaukee	Denver		
	St. Louis	Honolulu		
	Birmingham	Portland		
	Tampa-St. Petersburg			
	Atlanta			
	New Orleans			
	Nashville			
	Norfolk-Portsmouth			
	Dallas			
	Houston			
	San Antonio			

*Includes all metropolitan areas with estimated population of 500,000 or more in 1968.
SOURCE: Appendix B, Table 11.

TABLE 13
Estimated needs for new public comprehensive colleges and new public community colleges by 1980, by state

Region and state	Comprehensive colleges	Community colleges
New England states		
Connecticut	4-5	1-2
Maine		4-5
Massachusetts		6-8
New Hampshire		2-3
Rhode Island		2-3
Vermont		3-4
North Atlantic states		
Delaware	1	2-3
Maryland		3-4
New Jersey	3-4	5-8
New York	4-6	8-10
Pennsylvania	4-6	6-7
West Virginia		2-3
North Midwest states		
Illinois	4-5	4-6
Indiana	4	7-8
Michigan	3-4	6-7
Ohio	8	10-14
Wisconsin	1-2	1-2
Central states		
Iowa	3-4	3-4
Kansas	1	
Minnesota	2-3	3-4
Missouri	2-3	2-3
Nebraska	1	1-2
North Dakota		2-3
South Dakota		3-4
Southeast states		
Alabama		2-3
Arkansas	1	6-7
Florida	5-6	9-12
Georgia	1	4-5
Kentucky	2-3	2-3
Louisiana	1-2	3-4
Mississippi		2-3

TABLE 13 (continued)

Region and state	Comprehensive colleges	Community colleges
North Carolina		
South Carolina		
Tennessee	1	4–5
Virginia		2–4
Southwest states		
Arizona	4–5	4–5
New Mexico		3–4
Oklahoma	1	3–4
Texas	4	5–7
Mountain states		
Colorado	2–3	5–6
Idaho		3–4
Montana		2–3
Nevada		1–2
Utah	2–3	3–4
Wyoming		
Pacific states		
Alaska		
California	10–12	20–25
Hawaii	1	
Oregon	1	3–4
Washington	1–2	2–3
TOTAL	82–103	174–234

SOURCE: Estimates developed by Carnegie Commission staff.

References

1 Trow, M.: "The Second Transformation of American Secondary Education," *International Journal of Comparative Sociology,* vol. II, pp. 145–166, September 1961.

2 Greeley, A. M.: *From Backwater to Mainstream: A Profile of Catholic Higher Education,* McGraw-Hill Book Company, New York, 1969.

3 Keeton, M. T.: *Models and Mavericks: A Profile of Private Liberal Arts Colleges,* McGraw-Hill Book Company, New York, 1971.

4 Dunham, E. A.: *Colleges of the Forgotten Americans: A Profile of State Colleges and Regional Universities,* McGraw-Hill Book Company, New York, 1969.

5 California State Department of Education: *A Master Plan for Higher Education in California,* Sacramento, 1960.

6 Altman, R. A.: *The Upper Division College,* Jossey-Bass, Inc., San Francisco, 1970.

7 U.S. Office of Education: *Digest of Educational Statistics, 1969,* Washington, D.C., 1969.

8 Lee, E. C., and F. M. Bowen: *The Multicampus University: A Study of Academic Governance,* to be published by McGraw-Hill Book Company for the Carnegie Commission on Higher Education in 1971.

9 U.S. Bureau of the Census: "School Enrollment: October 1970," *Current Population Reports,* ser. P-20, no. 222, Washington, D.C., 1971.

10 U.S. Bureau of the Census: "Characteristics of American Youth: 1970," *Current Population Reports,* ser. P-23, no. 34, Washington, D.C., 1971.

11 U.S. Bureau of the Census: "The Social and Economic Status of Negroes in the United States, 1970," *Current Population Reports,* ser. P-23, no. 38, Washington, D.C., 1971.

12 U.S. Bureau of the Census: "Persons of Spanish Origin in the United States: November 1969," *Current Population Reports,* ser. P-20, no. 213, Washington, D.C., 1971.

13 U.S. Bureau of the Census: "Educational Attainment, March 1969," *Current Population Reports,* ser. P-20, no. 194, Washington, D.C., 1970.

14 *Chronicle of Higher Education,* April 12, 1967.

15 "One Year Follow-up Studies," *Project Talent,* American Institutes of Research, Pittsburgh, 1966.

16 Medsker, L. L., and J. W. Trent: *The Influence of Different Types of Public Higher Institutions on College Attendance from Varying Socioeconomic and Ability Levels,* Center for Research and Development in Higher Education, University of California, Berkeley, 1965.

17 SCOPE: *Four-State Profile,* grade 12, 1966, College Entrance Examination Board, New York, December 1966.

18 Withey, S., et al.: *A Degree and What Else?: Correlates and Consequences of a College Education,* to be published by McGraw-Hill Book Company for the Carnegie Commission on Higher Education in 1971.

19 *Chronicle of Higher Education,* March 29, 1971.

20 Carpenter, R., secretary of Smith College, quoted in *Chronicle of Higher Education,* March 29, 1971.

21 Jencks, C., and D. Riesman: *The Academic Revolution,* Doubleday & Company, Inc., Garden City, N.Y., 1968.

22 U.S. Office of Education: unpublished data on residence and migration, 1968.

23 Carnegie Commission on Higher Education: *The Capitol and the Campus: State Responsibility for Postsecondary Education,* McGraw-Hill Book Company, New York, 1971.

24 U.S. Office of Education: *Residence and Migration of Students: Basic State-to-State Matrix Tables,* Fall 1968, Washington, D.C., 1970.

25 *New York Times,* April 11, 1971.

26 *Chronicle of Higher Education,* May 4, 1970.

27 Trent, J. W., and L. L. Medsker: *Beyond High School,* Jossey-Bass, Inc., San Francisco, 1968.

28 Haggstrom, G. W.: *The Growth of Graduate Education in the Post-Sputnik Era,* to be published by the Carnegie Commission on Higher Education.

29 Carnegie Commission on Higher Education: *The Open-Door Colleges: Policies for Community Colleges,* McGraw-Hill Book Company, New York, 1970.

30 Berls, R. H.: "Higher Education Opportunity and Achievement in the United States," in *The Economics and Financing of Higher Education in the United States,* Joint Economic Committee, U.S. Congress, 1969.

31 Froomkin, J., and M. Pfeferman: "A Computer Model to Measure the Requirements for Student Aid in Higher Education," *Proceedings of the Social Statistics Section,* American Statistical Association, 1969.

32 **Carnegie Commission on Higher Education:** *Quality and Equality: Revised Recommendations, New Levels of Federal Responsibility for Higher Education,* McGraw-Hill Book Company, New York, 1970.

33 **U.S. Office of Education:** *Digest of Educational Statistics, 1970,* Washington, D.C., 1970.

34 **Trow, M.:** "Reflections on the Transition from Mass to Universal Higher Education," *Daedalus,* Winter 1970, pp. 1–42.

35 **Freeman, R. B.:** *The Market for College-Trained Manpower: A Study in the Economics of Career Choice,* Harvard University Press, Cambridge, Mass., 1971.

36 *Chronicle of Higher Education,* February 1, 1971.

37 *Chronicle of Higher Education,* March 8, 1971.

38 *Chronicle of Higher Education,* April 19, 1971.

39 *Chronicle of Higher Education,* October 26, 1970.

40 **Systems Research Group:** *Cost and Benefit Study of Post-Secondary Education in the Province of Ontario, School Year, 1968/1969,* University of Toronto, n.d. (Mimeographed.)

41 **Hodgkinson, H. L.:** *Institutions in Transition: A Study of Change in Higher Education,* Carnegie Commission on Higher Education, Berkeley, Calif., 1970.

42 **Peterson, R., and J. Bilorusky:** *May 1970: The Campus Aftermath of Cambodia and Kent State,* Carnegie Commission on Higher Education, Berkeley, Calif., 1971.

43 **Roose, K. D., and C. J. Anderson:** *A Rating of Graduate Programs,* American Council on Education, Washington, D.C., 1970.

44 **Southern Regional Education Board:** *Issues in Higher Education,* no. 2, 1971.

45 **Bowen, H. R., and G. K. Douglass:** *Efficiency in Liberal Education,* to be published by McGraw-Hill Book Company for the Carnegie Commission on Higher Education in 1971.

46 **Gaff, J. G., and Associates:** *The Cluster College,* Jossey-Bass., Inc., San Francisco, 1970.

47 **Singletary, O. E.** (ed.): *American Universities and Colleges,* 10th ed., American Council on Education, Washington, D.C., 1968.

48 **Moore, R. S.:** *A Guide to Higher Education Consortiums, 1965–66,* U.S. Office of Education, Washington, D.C., 1967.

49 Johnson, E. L.: "Consortia in Higher Education," *Educational Record,* Fall 1967.

50 Wood, H. H.: "College 'Common Markets' Growing," *New York Times,* January 11, 1971.

51 Moore, R. S., director, Center for Advanced Intercultural Studies, Chicago, Ill.: paper presented at the Presidents' Conference, Buck Hill Falls, Pa., December 4, 1967.

52 *Chronicle of Higher Education,* November 30, 1970.

53 Willingham, W. W.: *Educational Opportunity and the Organization of Higher Education,* Access Research Office, College Entrance Examination Board, Palo Alto, Calif., 1970.

54 Knowles, M. S.: *Higher Adult Education in the United States: Current Picture, Trends and Issues,* prepared for the Committee on Higher Adult Education, American Council on Education, Washington, D.C., 1969.

55 **The Times Education Supplement,** London, various issues.

56 Burkett, J. E., and P. Ruggiers (eds.): *Bachelor of Liberal Studies: Development of a Curriculum at the University of Oklahoma,* Center for the Study of Liberal Education for Adults at Boston University, Brookline, Mass., 1965.

57 Information supplied by Roy Troutt, dean, College of Liberal Studies, University of Oklahoma.

58 *Chronicle of Higher Education,* September 28, 1970.

59 *Report on Education Research,* vol. 3, p. 3, March 3, 1971.

60 *Chronicle of Higher Education,* April 26, 1971.

61 *State College and Bellefonte, Pa. Centre Times,* April 5, 1971.

62 **President's Task Force on the Extended University:** *Progress Report,* University of California, March 12, 1971.

63 *ECS Bulletin,* Education Commission of the States, Denver, Colo., February 1971.

64 *Chronicle of Higher Education,* December 14, 1970.

65 *New York Times,* May 24, 1971.

66 *Chronicle of Higher Education,* January 11, 1971.

67 *Chronicle of Higher Education,* May 3, 1971.

68 **Carnegie Commission on Higher Education:** *Less Time, More Options: Education Beyond the High School,* McGraw-Hill Book Company, New York, 1971.

Carnegie Commission on Higher Education
Sponsored Research Studies

A DEGREE AND WHAT ELSE?:
CORRELATES AND CONSEQUENCES OF A
COLLEGE EDUCATION
Stephen B. Withey, Jo Anne Coble, Gerald Gurin, John P. Robinson, Burkhard Strumpel, Elizabeth Keogh Taylor, and Arthur C. Wolfe

THE MULTICAMPUS UNIVERSITY:
A STUDY OF ACADEMIC GOVERNANCE
Eugene C. Lee and Frank M. Bowen

INSTITUTIONS IN TRANSITION:
A PROFILE OF CHANGE IN HIGHER
EDUCATION
(INCORPORATING THE 1970 STATISTICAL
REPORT)
Harold L. Hodgkinson

EFFICIENCY IN LIBERAL EDUCATION:
A STUDY OF COMPARATIVE INSTRUCTIONAL
COSTS FOR DIFFERENT WAYS OF ORGANIZ-
ING TEACHING-LEARNING IN A LIBERAL ARTS
COLLEGE
Howard R. Bowen and Gordon K. Douglass

CREDIT FOR COLLEGE:
PUBLIC POLICY FOR STUDENT LOANS
Robert W. Hartman

MODELS AND MAVERICKS:
A PROFILE OF PRIVATE LIBERAL ARTS
COLLEGES
Morris T. Keeton

BETWEEN TWO WORLDS:
A PROFILE OF NEGRO HIGHER EDUCATION
Frank Bowles and Frank A. DeCosta

BREAKING THE ACCESS BARRIERS:
A PROFILE OF TWO-YEAR COLLEGES
Leland L. Medsker and Dale Tillery

ANY PERSON, ANY STUDY:
AN ESSAY ON HIGHER EDUCATION IN THE
UNITED STATES
Eric Ashby

THE NEW DEPRESSION IN HIGHER
EDUCATION:
A STUDY OF FINANCIAL CONDITIONS AT 41
COLLEGES AND UNIVERSITIES
Earl F. Cheit

FINANCING MEDICAL EDUCATION:
AN ANALYSIS OF ALTERNATIVE POLICIES
AND MECHANISMS
Rashi Fein and Gerald I. Weber

HIGHER EDUCATION IN NINE COUNTRIES:
A COMPARATIVE STUDY OF COLLEGES AND
UNIVERSITIES ABROAD
Barbara B. Burn, Philip G. Altbach, Clark Kerr, and James A. Perkins

BRIDGES TO UNDERSTANDING:
INTERNATIONAL PROGRAMS OF AMERICAN
COLLEGES AND UNIVERSITIES
Irwin T. Sanders and Jennifer C. Ward

GRADUATE AND PROFESSIONAL EDUCATION,
1980:
A SURVEY OF INSTITUTIONAL PLANS
Lewis B. Mayhew

THE AMERICAN COLLEGE AND AMERICAN
CULTURE:
SOCIALIZATION AS A FUNCTION OF HIGHER
EDUCATION
Oscar and Mary F. Handlin

RECENT ALUMNI AND HIGHER EDUCATION:
A SURVEY OF COLLEGE GRADUATES
Joe L. Spaeth and Andrew M. Greeley

CHANGE IN EDUCATIONAL POLICY:
SELF-STUDIES IN SELECTED COLLEGES AND
UNIVERSITIES
Dwight R. Ladd

STATE OFFICIALS AND HIGHER EDUCATION:
A SURVEY OF THE OPINIONS AND
EXPECTATIONS OF POLICY MAKERS IN NINE
STATES
Heinz Eulau and Harold Quinley

ACADEMIC DEGREE STRUCTURES:
INNOVATIVE APPROACHES
PRINCIPLES OF REFORM IN DEGREE
STRUCTURES IN THE UNITED STATES
Stephen H. Spurr

COLLEGES OF THE FORGOTTEN AMERICANS:
A PROFILE OF STATE COLLEGES AND
REGIONAL UNIVERSITIES
E. Alden Dunham

FROM BACKWATER TO MAINSTREAM:
A PROFILE OF CATHOLIC HIGHER
EDUCATION
Andrew M. Greeley

THE ECONOMICS OF THE MAJOR PRIVATE
UNIVERSITIES
William G. Bowen
(Out of print, but available from University Microfilms.)

THE FINANCE OF HIGHER EDUCATION
Howard R. Bowen
(Out of print, but available from University Microfilms.)

ALTERNATIVE METHODS OF FEDERAL
FUNDING FOR HIGHER EDUCATION
Ron Wolk

INVENTORY OF CURRENT RESEARCH ON
HIGHER EDUCATION 1968
Dale M. Heckman and Warren Bryan Martin

The following reprints and technical reports are available from the Carnegie Commission on Higher Education, 1947 Center Street, Berkeley, California 94704.

TRENDS AND PROJECTIONS OF PHYSICIANS IN THE UNITED STATES 1967–2002, *by Mark S. Blumberg, published by Carnegie Commission, Berkeley, 1971 ($4.75).*

RESOURCE USE IN HIGHER EDUCATION: TRENDS IN OUTPUT AND INPUTS, 1930–1967, *by June O'Neill, published by Carnegie Commission, Berkeley, 1971 ($5.75).*

ACCELERATED PROGRAM OF MEDICAL EDUCATION, *by Mark S. Blumberg, reprinted from* JOURNAL OF MEDICAL EDUCATION, *vol. 46, no. 8, August 1971.*

SCIENTIFIC MANPOWER FOR 1970–1985, *by Allan M. Cartter, reprinted from* SCIENCE, *vol. 172, no. 3979, pp. 132–140, April 9, 1971.*

A NEW METHOD OF MEASURING STATES' HIGHER EDUCATION BURDEN, *by Neil Timm, reprinted from* THE JOURNAL OF HIGHER EDUCATION, *vol. 42, no. 1, pp. 27–33, January 1971.*

REGENT WATCHING, *by Earl F. Cheit, reprinted from* AGB REPORTS, *vol. 13, no. 6, pp. 4–13, March 1971.*

WHAT HAPPENS TO COLLEGE GENERATIONS POLITICALLY?, *by Seymour M. Lipset and Everett C. Ladd, Jr., reprinted from* THE PUBLIC INTEREST, *no. 24, Summer 1971.*

AMERICAN SOCIAL SCIENTISTS AND THE GROWTH OF CAMPUS POLITICAL ACTIVISM IN THE 1960s, by *Everett C. Ladd, Jr., and Seymour M. Lipset, reprinted from* SOCIAL SCIENCES INFORMATION, *vol. 10, no. 2, April 1971.*

THE POLITICS OF AMERICAN POLITICAL SCIENTISTS, by *Everett C. Ladd, Jr., and Seymour M. Lipset, reprinted from* PS, *vol. 4, no. 2, Spring 1971.*

THE DIVIDED PROFESSORIATE, by *Seymour M. Lipset and Everett C. Ladd, Jr., reprinted from* CHANGE, *vol. 3, no. 3, pp. 54–60, May 1971.*

JEWISH AND GENTILE ACADEMICS IN THE UNITED STATES: ACHIEVEMENTS, CULTURES AND POLITICS, by *Seymour M. Lipset and Everett C. Ladd, Jr., reprinted from* AMERICAN JEWISH YEAR BOOK, *1971.*

THE UNHOLY ALLIANCE AGAINST THE CAMPUS, by *Kenneth Keniston and Michael Lerner, reprinted from* NEW YORK TIMES MAGAZINE, *November 8, 1970.*

PRECARIOUS PROFESSORS: NEW PATTERNS OF REPRESENTATION, by *Joseph W. Garbarino, reprinted from* INDUSTRIAL RELATIONS, *vol. 10, no. 1, February 1971.*

... AND WHAT PROFESSORS THINK: ABOUT STUDENT PROTEST AND MANNERS, MORALS, POLITICS, AND CHAOS ON THE CAMPUS, by *Seymour Martin Lipset and Everett Carll Ladd, Jr., reprinted from* PSYCHOLOGY TODAY, *November 1970.**

DEMAND AND SUPPLY IN U.S. HIGHER EDUCATION: A PROGRESS REPORT, by *Roy Radner and Leonard S. Miller, reprinted from* AMERICAN ECONOMIC REVIEW, *May 1970.**

RESOURCES FOR HIGHER EDUCATION: AN ECONOMIST'S VIEW, by *Theodore W. Schultz, reprinted from* JOURNAL OF POLITICAL ECONOMY, *vol. 76, no. 3, University of Chicago, May/June 1968.**

INDUSTRIAL RELATIONS AND UNIVERSITY RELATIONS, by *Clark Kerr, reprinted from* PROCEEDINGS OF THE 21ST ANNUAL WINTER MEETING OF THE INDUSTRIAL RELATIONS RESEARCH ASSOCIATION, *pp, 15–25.**

NEW CHALLENGES TO THE COLLEGE AND UNIVERSITY, by *Clark Kerr, reprinted from Kermit Gordon (ed.),* AGENDA FOR THE NATION, *The Brookings Institution, Washington, D.C., 1968.**

PRESIDENTIAL DISCONTENT, by *Clark Kerr, reprinted from David C. Nichols (ed.),* PERSPECTIVES ON CAMPUS TENSIONS: PAPERS PREPARED FOR THE SPECIAL COMMITTEE ON CAMPUS TENSIONS, *American Council on Education, Washington, D.C., September 1970.**

STUDENT PROTEST—AN INSTITUTIONAL AND NATIONAL PROFILE, by *Harold Hodgkinson, reprinted from* THE RECORD, *vol. 71, no. 4, May 1970.**

WHAT'S BUGGING THE STUDENTS?, by *Kenneth Keniston, reprinted from* EDUCATIONAL RECORD, *American Council on Education, Washington, D.C., Spring 1970.**

THE POLITICS OF ACADEMIA, by *Seymour Martin Lipset, reprinted from David C. Nichols (ed.),* PERSPECTIVES ON CAMPUS TENSIONS: PAPERS PREPARED FOR THE SPECIAL COMMITTEE ON CAMPUS TENSIONS, *American Council on Education, Washington, D.C., September 1970.**

**The Commission's stock of this reprint has been exhausted.*